# THE
# MATHEMATICS
# OF GAMES

## JOHN D. BEASLEY

# DOVER PUBLICATIONS, INC.
## Mineola, New York

*Copyright*

Copyright © 1989, 2006 by John D. Beasley
All rights reserved.

*Bibliographical Note*

This Dover edition, first published in 2006, is an unabridged republication of
the 1990 Oxford paperback edition of the work originally published by Oxford
University Press, New York, in 1989. The author has provided a new Preface to
the Dover Edition.

*International Standard Book Number: 0-486-44976-9*

Manufactured in the United States of America
Dover Publications, Inc., 31 East 2nd Street, Mineola, N.Y. 11501

# PREFACE TO THE DOVER EDITION

This edition is a reprint rather than a revision and the text has not been altered, but sixteen years have passed since the book was written and two important developments should be acknowledged.

(a) A complete analysis of misère kayles (pages 118–119) has been published by William Sibert and John Conway, and the statement "the amount of work required appears to increase rather more than exponentially with $n$" at the top of page 119 has been found unduly pessimistic. For a convenient account, see *Winning Ways for Your Mathematical Plays* by E. R. Berlekamp, J. H. Conway, and R. K. Guy, second edition (A K Peters, Natick / Wellesley, Massachusetts, 2001–2004), volume 2, pages 446–452.

(b) It is reported that the game of driving the old woman to bed (pages 149–153) has been found by Marc Paulhus to contain positions that loop even when played with a standard fifty-two-card pack. See *Winning Ways for Your Mathematical Plays*, second edition, volume 4, page 892.

Additionally, I did R. P. Sprague an injustice in saying that "Grundy, apparently unlike Sprague, published follow-up work" (page 107). Two further papers by Sprague are cited in *Winning Ways for Your Mathematical Plays*, and two more are listed in *Games of No Chance* (ed. R. J. Nowakowski, Cambridge University Press, Cambridge, England, 1998) and in *More Games of No Chance* (ed. Nowakowski, Cambridge University Press, 2002). It remains true that "Grundy number" was the accepted usage in English when I was writing in 1989, but more recently I have seen the term "sprag" in print, and if I were writing today I think I would adopt this usage.

JOHN BEASLEY

*Harpenden, Hertfordshire, England*
*June 2005*

# ACKNOWLEDGEMENTS

Most of my debts to others are acknowledged in the body of the text, but some are more appropriately discharged here. David Wells read an early draft, and made many perceptive and sympathetic comments; the book is significantly richer as a result of his efforts. David Friedgood, Richard Guy, David Hooper, Terence Reese, John Roycroft, and Ken Whyld all gave helpful answers to enquiries, as did the Bodleian Library and the County Libraries of Hertfordshire (Harpenden) and West Sussex (Crawley, Shoreham-by-Sea, and Worthing). Sue helped with indexing and proofreading, and showed her customary forbearance in adjusting meals to my occasionally erratic requirements. And the staff of Oxford University Press showed their usual helpfulness and professionalism in the production of the book. My thanks to all.

Notwithstanding all this, if any error or inadequacy remains then it is my responsibility alone.

# CONTENTS

# I

# INTRODUCTION

The playing of games has long been a natural human leisure activity. References in art and literature go back for several thousand years, and archaeologists have uncovered many ancient objects which are most readily interpreted as gaming boards and pieces. The earliest games of all were probably races and other casual trials of strength, but games involving chance also appear to have a very long history. Figure 1.1 may well show such a game. Its rules have not survived, but other evidence supports the playing of dice games at this period.

**Figure 1.1** A wall-painting from an Egyptian tomb, *c*.2000 BC. The rules of the game have not survived, but the right hands of the players are clearly moving men on a board, while the left hands appear to have just rolled dice. From H. J. R. Murray, *A history of board games other than chess* (Oxford, 1952)

And if the playing of games is a natural instinct of all humans, the analysis of games is just as natural an instinct of mathematicians. Who should win? What is the best move? What are the odds of a certain chance event? How long is a game likely to take? When we are presented with a puzzle, are there standard techniques that will help us to find a solution? Does a particular puzzle have a solution at all? These are natural questions of mathematical interest, and we shall direct our attention to all of them.

To bring some order into our discussions, it is convenient to divide games into four classes:

(a) games of pure chance;

(b) games of mixed chance and skill;

(c) games of pure skill;

(d) automatic games.

There is a little overlap between these classes (for example, the children's game 'beggar your neighbour', which we shall treat as an automatic game, can also be regarded as a game of pure chance), but they provide a natural division of the mathematical ideas.

Our coverage of games of pure chance is in fact fairly brief, because the essentials will already be familiar to readers who have made even the most elementary study of the theory of probability. Nevertheless, the games cited in textbooks are often artificially simple, and there is room for an examination of real games as well. Chapters 2 and 3 therefore look at card and dice games respectively, and demonstrate some results which may be surprising. If, when designing a board for snakes and ladders, you want to place a snake so as to minimize a player's chance of climbing a particular ladder, where do you put it? Make a guess now, and then read Chapter 3; you will be in a very small minority if your guess proves to be right. These chapters also examine the effectiveness of various methods of randomization: shuffling cards, tossing coins, throwing dice, and generating allegedly 'random' numbers by computer.

Chapter 4 starts the discussion of games which depend both on chance and on skill. It considers the spread of results at ball games: golf (Figure 1.2), association football, and cricket. In theory, these are games of pure skill; in practice, they appear to contain a significant element of chance. The success of the player's stroke in Figure 1.2 will depend not only on how accurately he hits the ball but on how it negotiates any irregularities in the terrain. Some apparent chance influences on each of these games are examined, and it is seen to what extent they account for the observed spread of results.

Chapter 5 looks at ways of estimating the skill of a player. It considers both games such as golf, where each player returns an independent score, and chess, where a result merely indicates which of two players is the stronger. As an aside, it demonstrates situations in which the cyclic results '$A$ beats $B$, $B$ beats $C$, and $C$ beats $A$' may actually represent the normal expectation.

Chapter 6 looks at the determination of a player's optimal strategy in a game where one player knows something that the other does not.

**Figure 1.2** Golf: a drawing by C. A. Doyle entitled *Golf in Scotland* (from *London Society*, 1863). Play in a modern championship is more formalized, and urchins are no longer employed as caddies; but the underlying mathematical influences have not changed. *Mary Evans Picture Library*

This is the simplest case of the 'theory of games' of von Neumann. The value of bluffing in games such as poker is demonstrated, though no guarantee is given that the reader will become a millionaire as a result. The chapter also suggests some practical ways in which the players' chances in unbalanced games may be equalized.

Games of pure skill are considered in Chapters 7–10. Chapter 7 looks at puzzles, and demonstrates techniques both for solving them and for diagnosing those which are insoluble. Among the many puzzles considered are the 'fifteen' sliding block puzzle, the '*N* queens' puzzle both on a flat board and on a cylinder, Rubik's cube, peg solitaire, and the 'twelve coins' problem.

Chapter 8 examines 'impartial' games, in which the same moves are available to each player. It starts with the well-known game of nim, and shows how to diagnose and exploit a winning position. It then looks at some games which can be shown on examination to be clearly equivalent to nim, and it develops the remarkable theorem of Sprague and Grundy, according to which *every* impartial game whose rules guarantee termination is equivalent to nim.

Chapter 9 considers the relation between games and numbers. Much of the chapter is devoted to a version of nim in which each counter is owned by one player or the other; it shows how every pile of counters in such a game can be identified with a number, and how every number can be identified with a pile. This is the simplest case of the theory of 'numbers and games' which has recently been developed by Conway.

Chapter 10 completes the section on games of skill. It examines the concept of a 'hard' game; it looks at games in which it can be proved that a particular player can always force a win even though there may

**Figure 1.3** Chess: a drawing by J. P. Hasenclever (1810–53) entitled *The checkmate*. Perhaps White has been paying too much attention to his wine glass; at any rate, he has made an elementary blunder, and well deserves the guffaws of the spectators. *Mary Evans Picture Library*

be no realistic way of discovering how; and it discusses the paradox that a game of pure skill is playable only between players who are reasonably incompetent (Figure 1.3).

Finally, Chapter 11 looks at automatic games. These may seem mathematically trivial, but in fact they touch the deepest ground of all. It is shown that there is no general procedure for deciding whether an automatic game terminates, since a paradox would result if there were; and it is shown how this paradox throws light on the celebrated demonstration, by Kurt Gödel, that there are mathematical propositions which can be neither proved nor disproved.

Most of these topics are independent of each other, and readers with particular interests may freely select and skip. To avoid repetition, Chapters 4 and 5 refer to material in Chapters 2 and 3, but Chapter 6 stands on its own, and those whose primary interests are in games of pure skill can start anywhere from Chapter 7 onwards. Nevertheless, the analysis of games frequently brings pleasure in unexpected areas, and I hope that even those who have taken up the book with specific sections in mind will enjoy browsing through the remainder.

As regards the level of our mathematical treatment, little need be said. This is a book of results. Where a proof can easily be given in the normal course of exposition, it has been; where a proof is difficult or tedious, it has usually been omitted. However, there are proofs whose elegance, once comprehended, more than compensates for any initial difficulty; striking examples occur in Euler's analysis of the queens on a cylinder, Conway's of the solitaire army, and Hutchings's of the game now known as 'sylver coinage'. These analyses have been included in full, even though they are a little above the general level of the book. If you are looking only for light reading, you can skip them, but I hope that you will not; they are among my favourite pieces of mathematics, and I shall be surprised if they do not become among yours as well.

# 2

# THE LUCK OF THE DEAL

This chapter looks at some of the probabilities governing play with cards, and examines the effectiveness of practical shuffling.

## Counting made easy

Most probabilities relating to card games can be determined by counting. We count the total number of possible hands, and the number having some desired property. The ratio of these two numbers gives the probability that a hand chosen at random does indeed have the desired property.

The counting can often be simplified by making use of a well-known formula: if we have $n$ things, we can select a subset of $r$ of them in $n!/\{r!(n-r)!\}$ different ways, where $n!$ stands for the repeated product $n \times (n-1) \times \ldots \times 1$. This is easily proved. We can arrange $r$ things in $r!$ different ways, since we can choose any of them to be first, any of the remaining $(r-1)$ to be second, any of the $(r-2)$ still remaining to be third, and so on. Similarly, we can arrange $r$ things out of $n$ in $n \times (n-1) \times \ldots \times (n-r+1)$ different ways, since we can choose any of them to be first, any of the remaining $(n-1)$ to be second, any of the $(n-2)$ still remaining to be third, and so on down to the $(n-r+1)$th; and this product is clearly $n!/(n-r)!$. But this gives us every possible *arrangement* of $r$ things out of $n$, and we must divide by $r!$ to get the number of selections of $r$ things that are actually different.

The formula $n!/\{r!(n-r)!\}$ is usually denoted by $\binom{n}{r}$. The derivation above applies only if $1 \leqslant r \leqslant (n-1)$, but the formula can be extended to cover the whole range $0 \leqslant r \leqslant n$ by defining 0! to be 1. Its values form the well-known 'Pascal's Triangle'. For $n \leqslant 10$, they are shown in Table 2.1.

**Table 2.1** The first ten rows of Pascal's triangle

| | *r* | | | | | | | | | | |
|---|---|---|---|---|---|---|---|---|---|---|---|
| *n* | 0 | 1 | 2 | 3 | 4 | 5 | 6 | 7 | 8 | 9 | 10 |
| 0 | 1 | | | | | | | | | | |
| 1 | 1 | 1 | | | | | | | | | |
| 2 | 1 | 2 | 1 | | | | | | | | |
| 3 | 1 | 3 | 3 | 1 | | | | | | | |
| 4 | 1 | 4 | 6 | 4 | 1 | | | | | | |
| 5 | 1 | 5 | 10 | 10 | 5 | 1 | | | | | |
| 6 | 1 | 6 | 15 | 20 | 15 | 6 | 1 | | | | |
| 7 | 1 | 7 | 21 | 35 | 35 | 21 | 7 | 1 | | | |
| 8 | 1 | 8 | 28 | 56 | 70 | 56 | 28 | 8 | 1 | | |
| 9 | 1 | 9 | 36 | 84 | 126 | 126 | 84 | 36 | 9 | 1 | |
| 10 | 1 | 10 | 45 | 120 | 210 | 252 | 210 | 120 | 45 | 10 | 1 |

The function tabulated is $\binom{n}{r}=n!/\{r!(n-r)!\}$.

To see how this formula works, let us look at some distributional problems at bridge and whist. In these games, the pack consists of four 13-card suits (spades, hearts, diamonds, and clubs), each player receives thirteen cards in the deal, and opposite players play as partners. For example, if we have a hand containing four spades, what is the probability that our partner has at least four spades also?

First, let us count the total number of different hands that partner may hold. We hold 13 cards ourselves, so partner's hand must be taken from the remaining 39; but it must contain 13 cards, and we have just seen that the number of different selections of 13 cards that can be made from 39 is $\binom{39}{13}$. So this is the number of different hands that partner may hold. It does not appear in Table 2.1, but it can be calculated as $(39\times 38\times \ldots \times 27)/(13\times 12\times \ldots \times 1)$, and it amounts to 8 122 425 444. Such numbers are usually rounded off to a sensible number of decimals ($8.12\times 10^9$, for example), but it is convenient to work out this first example in full.

Now let us count the number of hands in which partner holds precisely four spades. Nine spades are available to him, so he has $\binom{9}{4}=126$ possible spade holdings. Similarly, 30 non-spades are available to him, and his hand must contain nine non-spades, so he has $\binom{30}{9}=14\ 307\ 150$ possible non-spade holdings. Each spade holding can be married with each of the non-spade holdings, which gives 1 802 700 900 possible hands containing precisely four spades.

So, if partner's hand has been dealt at random from the 39 cards available to him, the probability that he has exactly four spades is given by dividing the number of hands containing exactly four spades (1 802 700 900) by the total number of possible hands (8 122 425 444). This gives 0.222 to three decimal places.

A similar calculation can be performed for each number of spades that partner can hold, and summation of the results gives the figures shown in the first column of Table 2.2. In particular, we see that if we hold four spades ourselves, the probability that partner holds at least four more is approximately 0.34, or only a little better than one third. The remainder of Table 2.2 has been calculated in the same way, and shows the probability that partner holds at least a certain number of spades, given that we ourselves hold five or more.[1]

**Table 2.2**  Bridge: the probabilities of partner's suit holdings

| Partner's holding | Player's own holding | | | | | | | | |
|---|---|---|---|---|---|---|---|---|---|
| | 4 | 5 | 6 | 7 | 8 | 9 | 10 | 11 | 12 |
| 1+ | 0.99 | 0.97 | 0.96 | 0.93 | 0.89 | 0.82 | 0.72 | 0.56 | 0.33 |
| 2+ | 0.89 | 0.84 | 0.76 | 0.67 | 0.55 | 0.41 | 0.25 | 0.11 | |
| 3+ | 0.65 | 0.54 | 0.43 | 0.31 | 0.20 | 0.10 | 0.03 | | |
| 4+ | 0.34 | 0.24 | 0.15 | 0.08 | 0.03 | 0.01 | | | |
| 5+ | 0.11 | 0.06 | 0.03 | 0.01 | 0.00 | | | | |
| 6+ | 0.02 | 0.01 | 0.00 | 0.00 | | | | | |
| 7+ | 0.00 | 0.00 | 0.00 | | | | | | |

[1] It is not the purpose of this book to give advice on specific games, but I cannot resist pointing out an implication that is sometimes overlooked. Suppose that your partner at bridge has opened the bidding with 'one no trump', indicating a balanced hand of some agreed strength, and that you yourself hold a four-card major suit and enough all-round strength to bid game: the sort of hand on which you want to be in four of the major if partner also holds four, and otherwise to be in 3NT. It is quite a common situation, and several artificial bidding conventions have been invented to deal with it. But Table 2.2 suggests that *only about one-third of the time will partner actually have the four-card fit that you seek;* and while the calculation of this table took into account unbalanced hands on which partner would not have opened 1NT, a revised calculation omitting such hands produces much the same answer. The remaining two-thirds of the time, you will end up in 3NT anyway, and all the bidding convention will have done is to pinpoint a probable weakness in declarer's hand; against competent opponents, you would actually have given your side a better practical chance by bidding an immediate 'three no nonsense' and leaving the defenders to guess. Of course, this simple calculation cannot say whether the gains when partner does have a fit are likely to outweigh the losses when he does not, but it is instructive that the latter case occurs twice as often as the former.

Table 2.3 shows the other side of the coin. It assumes that our side holds a certain number of cards in a suit, and shows how the remaining cards are likely to be distributed between our opponents. Note that the most even distribution is not necessarily the most probable. For example, if we hold seven cards of a suit ourselves, the remaining six are more likely to be distributed 4-2 than 3-3, because '4-2' actually covers the two cases 4-2 and 2-4.

**Table 2.3** Bridge: the probabilities of opponents' suit holdings

| Number of cards held by the partnership | | | | | | |
|---|---|---|---|---|---|---|
| 5 | 6 | 7 | 8 | 9 | 10 | 11 |
| 4-4 0.33 | 4-3 0.62 | 3-3 0.36 | 3-2 0.68 | 2-2 0.41 | 2-1 0.78 | 1-1 0.52 |
| 5-3 0.47 | 5-2 0.31 | 4-2 0.48 | 4-1 0.28 | 3-1 0.50 | 3-0 0.22 | 2-0 0.48 |
| 6-2 0.17 | 6-1 0.07 | 5-1 0.15 | 5-0 0.04 | 4-0 0.10 | | |
| 7-1 0.03 | 7-0 0.01 | 6-0 0.01 | | | | |
| 8-0 0.00 | | | | | | |

Tables 2.2 and 2.3 do not constitute a complete recipe for success at bridge, because the bidding and play may well give reason to suspect abnormal distributions, but they provide a good foundation.

# 4-3-3-3 and all that

A similar technique can be used to determine the probabilities of the possible suit distributions in a hand.

For example, let us work out the probability of the distribution 4-3-3-3 (four cards in one suit and three in each of the others). Suppose for a moment that the four-card suit is spades. There are now $\binom{13}{4}$ possible spade holdings and $\binom{13}{3}$ possible holdings in each of the other three suits, and any combination of these can be joined together to give a 4-3-3-3 hand with four spades. Additionally, the four-card suit may be chosen in four ways, so the total number of 4-3-3-3 hands is $4 \times \binom{13}{4} \times \binom{13}{3} \times \binom{13}{3} \times \binom{13}{3}$. But the total number of 13-card hands that can be dealt from a 52-card pack is $\binom{52}{13}$, so the probability of a 4-3-3-3 hand is $4 \times \binom{13}{4} \times \binom{13}{3} \times \binom{13}{3} \times \binom{13}{3} / \binom{52}{13}$. This works out to 0.105 to three decimal places.

Equivalent calculations can be performed for other distributions. In the case of a distribution which has only two equal suits, such as

4-4-3-2, there are twelve ways in which the distribution can be achieved (4S-4H-3D-2C, 4S-4H-2D-3C, 4S-3H-4D-2C, 4S-3H-2D-4C, and so on), while in the case of a distribution with no equal suits, such as 5-4-3-1, there are 24 ways. The multiplying factor 4 must therefore be replaced by 12 or 24 as appropriate. This leads to the same effect that we saw in Table 2.3: the most even distribution (4-3-3-3) is by no means the most probable. Indeed, it is only the fifth most probable, coming behind 4-4-3-2, 5-3-3-2, 5-4-3-1, and 5-4-2-2.

The probabilities of all the possible distributions are shown in Table 2.4. This table allows some interesting conclusions to be drawn. For example, over 20 per cent of all hands contain a suit of at least six cards, and 4 per cent contain a suit of at least seven; over 35 per cent contain a very short suit (singleton or void), and 5 per cent contain an actual void. These probabilities are perhaps rather higher than might have been guessed before the calculation was performed. Note also, as a curiosity, that the probabilities of 7-5-1-0 and 8-3-2-0 are exactly equal. This may be verified by writing out their factorial expressions in full and comparing them.

**Table 2.4** Bridge: the probabilities of suit distributions

| Distribution | Probability | Distribution | Probability |
|---|---|---|---|
| 4-4-3-2 | 0.216 ($2.2 \times 10^{-1}$) | 8-2-2-1 | 0.002 ($1.9 \times 10^{-3}$) |
| 5-3-3-2 | 0.155 ($1.6 \times 10^{-1}$) | 8-3-1-1 | 0.001 ($1.2 \times 10^{-3}$) |
| 5-4-3-1 | 0.129 ($1.3 \times 10^{-1}$) | 7-5-1-0 | 0.001 ($1.1 \times 10^{-3}$) |
| 5-4-2-2 | 0.106 ($1.1 \times 10^{-1}$) | 8-3-2-0 | 0.001 ($1.1 \times 10^{-3}$) |
| 4-3-3-3 | 0.105 ($1.1 \times 10^{-1}$) | 6-6-1-0 | 0.001 ($7.2 \times 10^{-4}$) |
| 6-3-2-2 | 0.056 ($5.6 \times 10^{-2}$) | 8-4-1-0 | 0.000 ($4.5 \times 10^{-4}$) |
| 6-4-2-1 | 0.047 ($4.7 \times 10^{-2}$) | 9-2-1-1 | 0.000 ($1.8 \times 10^{-4}$) |
| 6-3-3-1 | 0.034 ($3.4 \times 10^{-2}$) | 9-3-1-0 | 0.000 ($1.0 \times 10^{-4}$) |
| 5-5-2-1 | 0.032 ($3.2 \times 10^{-2}$) | 9-2-2-0 | 0.000 ($8.2 \times 10^{-5}$) |
| 4-4-4-1 | 0.030 ($3.0 \times 10^{-2}$) | 7-6-0-0 | 0.000 ($5.6 \times 10^{-5}$) |
| 7-3-2-1 | 0.019 ($1.9 \times 10^{-2}$) | 8-5-0-0 | 0.000 ($3.1 \times 10^{-5}$) |
| 6-4-3-0 | 0.013 ($1.3 \times 10^{-2}$) | 10-2-1-0 | 0.000 ($1.1 \times 10^{-5}$) |
| 5-4-4-0 | 0.012 ($1.2 \times 10^{-2}$) | 9-4-0-0 | 0.000 ($9.7 \times 10^{-6}$) |
| 5-5-3-0 | 0.009 ($9.0 \times 10^{-3}$) | 10-1-1-1 | 0.000 ($4.0 \times 10^{-6}$) |
| 6-5-1-1 | 0.007 ($7.1 \times 10^{-3}$) | 10-3-0-0 | 0.000 ($1.5 \times 10^{-6}$) |
| 6-5-2-0 | 0.007 ($6.5 \times 10^{-3}$) | 11-1-1-0 | 0.000 ($2.5 \times 10^{-7}$) |
| 7-2-2-2 | 0.005 ($5.1 \times 10^{-3}$) | 11-2-0-0 | 0.000 ($1.1 \times 10^{-7}$) |
| 7-4-1-1 | 0.004 ($3.9 \times 10^{-3}$) | 12-1-0-0 | 0.000 ($3.2 \times 10^{-9}$) |
| 7-4-2-0 | 0.004 ($3.6 \times 10^{-3}$) | 13-0-0-0 | 0.000 ($6.3 \times 10^{-12}$) |
| 7-3-3-0 | 0.003 ($2.7 \times 10^{-3}$) | | |

Table 2.4 shows that the probability of a 13-0-0-0 distribution is approximately $6.3 \times 10^{-12}$. The probability that all four hands have this distribution can be calculated similarly, and proves to be approximately $4.5 \times 10^{-28}$. Now if an event has a very small probability $p$, it is necessary to perform approximately $0.7/p$ trials in order to obtain an even chance of its occurrence.[2] A typical evening's bridge comprises perhaps twenty deals, so a once-a-week player must play for over one hundred million years to have an even chance of receiving a thirteen-card suit. If ten million players are active once a week, a hand containing a thirteen-card suit may be expected about once every fifteen years, but it is still extremely unlikely that a genuine deal will produce four such hands.

# Shuffle the pack and deal again

So far, we have assumed the cards to have been dealt at random, each of the possible distributions being equally likely irrespective of the previous history of the pack. The way in which previous history is destroyed in practice is by shuffling, so it is appropriate to have a brief look at this process.

Let us restrict the pack to six cards for a moment, and let us consider the shuffle shown in Figure 2.1. The card which was in position 1 has moved to position 4, that which was in position 4 has moved to position 6, and that which was in position 6 has moved to position 1. So the cards in positions 1, 4, and 6 have cycled among themselves. We call this movement a *three-cycle*, and we denote it by (1,4,6). Similarly, the cards in positions 2 and 5 have interchanged places, which can be represented by the two-cycle (2,5), and the card in position 3 has stayed put, which can be represented by the one-cycle (3). So the complete shuffle is represented by these three cycles. We call this *representation by disjoint cycles*, the word 'disjoint' signifying that no two cycles involve a common card. A similar representation can be obtained for any shuffle of a pack of any size. If the pack contains $n$ cards, a particular shuffle may be represented by anything from a single $n$-cycle to $n$ separate one-cycles.

Now let us suppose for a moment that our shuffle can be represented by the single $k$-cycle $(A,B,C,\ldots,K)$, and let us consider the card at position $A$. If we perform the shuffle once, we move this card to

---

[2] This is a consequence of the Poisson distribution, which we shall meet in Chapter 4.

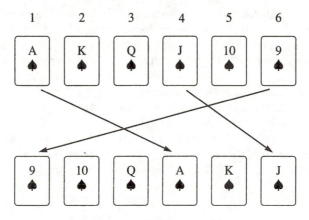

**Figure 2.1** A shuffle of six cards

position $B$; if we perform the same shuffle again, we move it on to $C$; and if we perform the shuffle a further $(k-2)$ times, we move it the rest of the way round the cycle and back to $A$. The same is plainly true of the other cards in the cycle, so the performance of a $k$-cycle $k$ times moves every card back to its original position. More generally, if the representation of a shuffle by disjoint cycles consists of cycles of lengths $a, b, \ldots, m$, and if $p$ is the lowest common multiple (LCM) of $a, b, \ldots, m$, then the performance of the shuffle $p$ times moves every card back to its starting position. We call this value $p$ the *period* of the shuffle.

So if a pack contains $n$ cards, the longest period that a shuffle can have may be found by considering all the possible partitions of $n$ and choosing that with the greatest LCM. For packs of reasonable size, this is not difficult, and the longest periods of shuffles of all packs not exceeding 52 cards are shown in Table 2.5. In particular, the longest period of a shuffle of 52 cards is 180 180, this being the period of a shuffle containing cycles of lengths 4, 5, 7, 9, 11, and 13. These six cycles involve only 49 cards, but no longer cycle can be obtained by involving the three remaining cards as well; for example, replacing the 13-cycle by a 16-cycle would actually reduce the period (by a factor 13/4). So all that we can do with the odd three cards is to permute them among themselves by a three-cycle, a two-cycle and a one-cycle, or three one-cycles, and none of these affects the period of the shuffle. According to Martin Gardner, this calculation seems first

**Table 2.5** The shuffles of longest period

| Pack size | Longest period | Component cycles |
|---|---|---|
| 2 | 2 | 2 |
| 3 | 3 | 3 |
| 4 | 4 | 4 |
| 5–6 | 6 | 2, 3 |
| 7 | 12 | 3, 4 |
| 8 | 15 | 3, 5 |
| 9 | 20 | 4, 5 |
| 10–11 | 30 | 2, 3, 5 |
| 12–13 | 60 | 3, 4, 5 |
| 14 | 84 | 3, 4, 7 |
| 15 | 105 | 3, 5, 7 |
| 16 | 140 | 4, 5, 7 |
| 17–18 | 210 | 2, 3, 5, 7 |
| 19–22 | 420 | 3, 4, 5, 7 |
| 23–24 | 840 | 3, 5, 7, 8 |
| 25–26 | 1260 | 4, 5, 7, 9 |
| 27 | 1540 | 4, 5, 7, 11 |
| 28 | 2310 | 2, 3, 5, 7, 11 |
| 29 | 2520 | 5, 7, 8, 9 |
| 30–31 | 4620 | 3, 4, 5, 7, 11 |
| 32–33 | 5460 | 3, 4, 5, 7, 13 |
| 34–35 | 9240 | 3, 5, 7, 8, 11 |
| 36–37 | 13860 | 4, 5, 7, 9, 11 |
| 38–39 | 16380 | 4, 5, 7, 9, 13 |
| 40 | 27720 | 5, 7, 8, 9, 11 |
| 41 | 30030 | 2, 3, 5, 7, 11, 13 |
| 42 | 32760 | 5, 7, 8, 9, 13 |
| 43–46 | 60060 | 3, 4, 5, 7, 11, 13 |
| 47–48 | 120120 | 3, 5, 7, 8, 11, 13 |
| 49–52 | 180180 | 4, 5, 7, 9, 11, 13 |

to have been performed by W. H. H. Hudson in 1865 (*Educational Times Reprints* **2** 105).

But 180 180 is merely the longest period that a shuffle of a 52-card pack can possess, and it does not follow that performing a particular shuffle 180 180 times brings the pack back to its original order. For example, one of the riffle shuffles which we shall consider later in the chapter has period 8. The number 8 is not a factor of 180 180, so performing this shuffle 180 180 times does *not* restore the pack to its original order. To guarantee that a 52-card pack is restored by repetiton of a shuffle, we must perform it $P$ times, where $P$ is a common multiple of all the numbers from 1 to 52 inclusive. The lowest such multiple is

$$2^5 \times 3^3 \times 5^2 \times 7^2 \times 11 \times 13 \times 17 \times 19 \times 23 \times 29 \times 31 \times 37 \times 41 \times 43 \times 47,$$

which works out to 3 099 044 504 245 996 706 400.

This is all very well, and not without interest, but the last thing that we want from a practical shuffle is a guarantee that we shall get back where we started. Fortunately, it is so difficult to repeat a shuffle exactly that such a guarantee would be almost worthless even if we were able to shuffle for long enough to bring it into effect. Nevertheless, what can we expect in practice?

If we do not shuffle at all, the new deal exactly reflects the play resulting from the previous deal. In a game such as bridge or whist, for example, the play consists of 'tricks'; one player leads a card, and each of the other players adds a card which must be of the same suit as the leader's if possible. The resulting set of four cards is collected and turned over before further play. A pack obtained by stacking such tricks therefore has a large amount of order, in that sets of four adjacent cards are much more likely to be from the same suit than would be the case if the pack were arranged at random. If we deal such a pack without shuffling, the distribution of the suits around the hands will be much more even than would be expected from a random deal.

A simple cut (Figure 2.2) merely cycles the hands among the players; it does not otherwise affect the distribution.

**Figure 2.2** A cut

Overhand shuffles (Figure 2.3) move the cards in blocks. They therefore break up the ordering to some extent, but only a few adjacencies are changed, and the resulting deals are still somewhat more likely to produce even distributions than a random deal. Overhand shuffles also provide the justification for the bridge player's rule that if no better guide is to hand then a declarer should play for a hidden queen to lie over the jack; if the queen covered the jack in the play of the previous hand, overhand shuffling may well have failed to separate them.

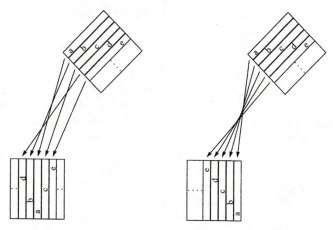

**Figure 2.3** Overhand shuffles

Riffle shuffles (Figure 2.4) behave quite differently. If performed perfectly, the pack being divided into two exact halves which are then interleaved, they change all the adjacencies but produce a pack in which two cards *separated by precisely one other* have the properties originally possessed by adjacent cards. If the same riffle is performed again, cards separated by precisely *three* others have these properties. If we stack thirteen spades on top of the pack, perform one perfect riffle shuffle, and deal, two partners get all the spades between them. If we do the same thing with two riffles, one player gets them all to himself. If we do it with three riffles, the spades are again divided between two players, but this time the players are opponents. What happens with four or more riffles depends on whether they are 'out' riffles (Figure 2.4 left, the pack being reordered 1,27,2,28,...,26,52) or 'in' riffles (Figure 2.4 right, the reordering of the pack now being 27,1,28,2,...,52,26). The 'out' riffle has period 8, and produces

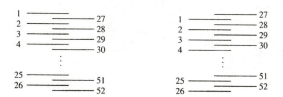

**Figure 2.4** Riffle shuffles

4-3-3-3 distributions of the spades if between four and eight riffles are used. The 'in' riffle has period 52 and produces somewhat more uneven distributions after the fifth riffle, but 26 such riffles completely reverse the order of the pack.

Fortunately, it is quite difficult to perform a perfect riffle shuffle. Expert card manipulators can do it, but such people are not usually found in friendly games; perhaps it is as well. It is nevertheless clear that small numbers of riffles may produce markedly abnormal distributions. In sufficiently expert hands, they may even provide a practicable way of obtaining a 'perfect deal' which delivers a complete suit to each player. New packs from certain manufacturers are usually arranged in suits, and if a card magician suspects such an arrangement, he can unobtrusively apply two perfect riffles to a pack *which has just been innocently bought by someone else* and present it for cutting and dealing. If the pack proves not to have been arranged in suits, or if somebody spoils the trick by shuffling the pack separately, the magician keeps quiet, and nobody need be any the wiser; but if the deal materializes as intended, everyone is at least temporarily amazed.

From the point of view of practical play, however, the unsatisfactory behaviour of the overhand and riffle shuffles suggests that every shuffle should include a thorough face-down mixing of cards on the table. If local custom frowns on this, so be it, but shuffles performed in the hand are unlikely to be fully effective.

# 3

# THE LUCK OF THE DIE

---

Other than cards, the most common media for controlling games of chance are dice. We look at some of their properties in this chapter.

## Counting again made easy

Although most modern dice are cubic, many other forms of dice exist. Prismatic dice have long been used (see for example R. C. Bell, *Board and table games from many civilisations*, Oxford, 1960, or consider the rolling of hexagonal pencils by schoolboys); so have teetotums (spinning polygonal tops) and sectioned wheels; so have dodecahedra and other regular solids; and so have computer-generated pseudo-random numbers. Examples of all except the last are shown in Figure 3.1. The simplest dice of all are two-sided, and can be obtained from banks and other gambling supply houses. We start by considering such a die, and we assume initially that each outcome has a probability of exactly one half.

We now examine the fundamental problem: If we toss a coin $n$ times, what is the probability that we obtain exactly $r$ heads?

If we toss once, there are only two possible outcomes, as shown in Figure 3.2 (upper left). Each of these outcomes has probability 1/2.

If we toss twice, there are four possible outcomes, as shown in Figure 3.2 (lower left). Each of these outcomes has probability 1/4. The numbers of outcomes containing 0, 1, and 2 heads are 1, 2, and 1 respectively, so the probabilities of obtaining these numbers of heads are 1/4, 1/2, and 1/4 respectively.

If we toss three times, there are eight possible outcomes, as shown in Figure 3.2 (right). Each of these outcomes has probability 1/8. The numbers of outcomes containing 0, 1, 2, and 3 heads are 1, 3, 3, and 1 respectively, so the probabilities of obtaining these numbers of heads are 1/8, 3/8, 3/8, and 1/8 respectively.

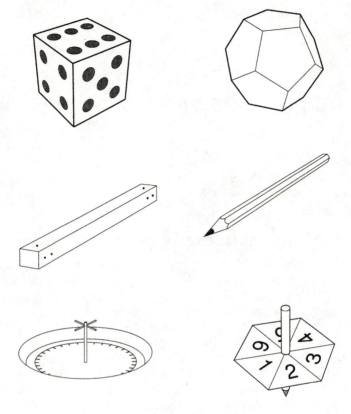

**Figure 3.1** Some typical dice

These results exemplify a general pattern. If we toss $n$ times, there are $2^n$ possible outcomes. These range from $TT\ldots T$ to $HH\ldots H$, and each has probability $1/2^n$. However, the number of outcomes which contain exactly $r$ heads is equal to the number of ways of selecting $r$ things from a set of $n$, and we saw in the last chapter that this is $\binom{n}{r}$. So the probability of obtaining exactly $r$ heads is $\binom{n}{r}/2^n$.

Now let us briefly consider a two-sided die in which the probabilities of the outcomes are unequal. Such dice are not unknown in practical play; Bell cites the use of cowrie shells, and the first innings in casual games of cricket during my boyhood was usually determined by the spinning of a bat. The probability of a 'head' (or of a cowrie shell falling with mouth upwards, or a bat falling on its face) is now some

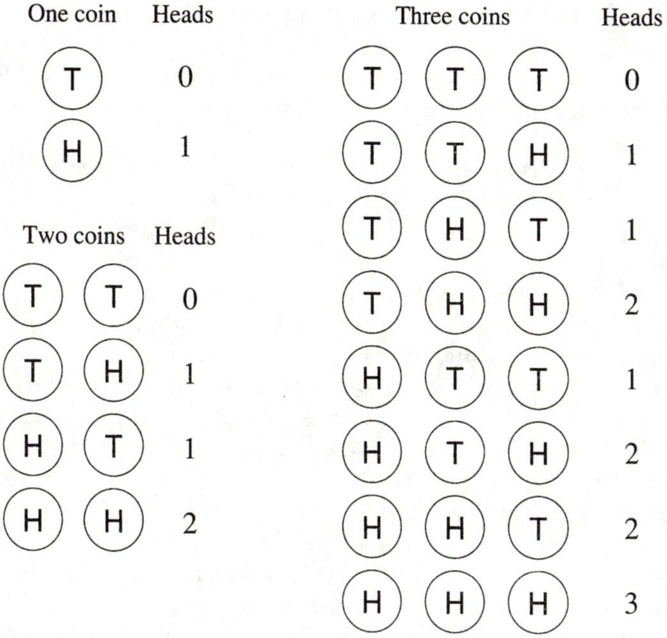

| One coin | Heads | Three coins | | | Heads |
|:---:|:---:|:---:|:---:|:---:|:---:|

**Figure 3.2** Tossing a coin

number $p$ rather than $1/2$, and the probability of a 'tail' is accordingly $(1-p)$. Now suppose that a particular sequence $HTHT \ldots$ of $n$ tosses contains $r$ heads. The probability of each head is $p$ and that of each tail is $(1-p)$, so the probability that we obtain precisely this sequence is $p^r(1-p)^{n-r}$. But there are $\binom{n}{r}$ sequences of $n$ tosses which contain $r$ heads, so the probability that we obtain exactly $r$ heads from $n$ tosses is $\binom{n}{r}p^r(1-p)^{n-r}$.

This important distribution is known as the *binomial distribution*. We shall meet it again in Chapter 4.

# The true law of averages

Mathematicians know that there is no such thing as the popular 'law of averages'. Events determined by chance do *not* remember previous results so as to even themselves out. For example, if we toss a coin a hundred times, we cannot expect always to get exactly fifty heads and fifty tails. What can we expect instead?

It is proved in textbooks on statistics that for large even $n$, the probability of getting exactly $n/2$ heads from $n$ tosses is approximately $\sqrt{(2/n\pi)}$. This approximation always overestimates the true probability, but the relative error is only about one part in $4n$. This is good enough for most purposes.

But of greater interest than the number of exactly even results is the spread around this point. This is conveniently measured by the standard deviation.[1] To calculate it from the definition becomes tedious as $n$ becomes large, but there are two convenient short cuts. The first is to appeal to a theorem relating to the binomial distribution, which states that the standard deviation of such a distribution is $\sqrt{\{np(1-p)\}}$. In the present case, $p = 1/2$, so the standard deviation is $\sqrt{(n/4)}$. The second is to appeal to an important general theorem known as the Central Limit Theorem, which tells us rather more about the distribution than just its standard deviation.

The Central Limit Theorem states that if we take a repeated sample from *any* population with mean $m$ and standard deviation $s$, the sum of the sample approaches a distribution known as the 'normal' distribution with mean $mn$ and standard deviation $s\sqrt{n}$. The normal distribution $N(x)$ with mean 0 and standard deviation 1 is shown in Table 3.1, and if we have a number from a normal distribution with mean $mn$ and standard deviation $s\sqrt{n}$, the probability that it is less than a particular value $y$ can be obtained by setting $x$ to $(y-mn)/s\sqrt{n}$ and looking up $N(x)$ in the table.

Table 3.1 is very useful as an estimator of coin tosses; provided that we toss at least eight times, it tells us the probability of obtaining a number of heads within any given range with an error not exceeding 0.01. Suppose that we want to find the approximate probability that tossing a coin a hundred times produces at least forty heads. The act of tossing a coin is equivalent to taking a sample from a population comprising the numbers 0 (tail) and 1 (head) in equal proportions. The mean of such a population is clearly $1/2$, and the deviation of every member from this mean is $\pm 1/2$, so the standard deviation is $1/2$. Tossing a hundred coins and counting the heads is equivalent to taking a sample of a hundred numbers and computing the sum, so if we want at least 40 heads (which means at least 39.5, since heads

---

[1] The standard deviation of a set is the root mean square deviation about the mean. If a set of $n$ numbers $x_1 \ldots x_n$ has mean $m$, its standard deviation $s$ is given by the formula $s = \sqrt{(\{(x_1-m)^2 + \ldots + (x_n-m)^2\}/n)}$. The standard deviation does not tell us everything about the distribution of a set of numbers; in particular, it does not tell us whether there are more values on one side of the mean than on the other. Nevertheless, it is a useful measure of the general spread of a set.

**Table 3.1** The standard normal distribution

| $x$ | $N(x)$ | $x$ | $N(x)$ | $x$ | $N(x)$ | $x$ | $N(x)$ |
|------|--------|------|--------|------|--------|------|--------|
| $-3.0$ | 0.001 | $-1.5$ | 0.067 | 0.0 | 0.500 | 1.5 | 0.933 |
| $-2.9$ | 0.002 | $-1.4$ | 0.081 | 0.1 | 0.540 | 1.6 | 0.945 |
| $-2.8$ | 0.003 | $-1.3$ | 0.097 | 0.2 | 0.579 | 1.7 | 0.955 |
| $-2.7$ | 0.003 | $-1.2$ | 0.115 | 0.3 | 0.618 | 1.8 | 0.964 |
| $-2.6$ | 0.005 | $-1.1$ | 0.136 | 0.4 | 0.655 | 1.9 | 0.971 |
| $-2.5$ | 0.006 | $-1.0$ | 0.159 | 0.5 | 0.691 | 2.0 | 0.977 |
| $-2.4$ | 0.008 | $-0.9$ | 0.184 | 0.6 | 0.726 | 2.1 | 0.982 |
| $-2.3$ | 0.011 | $-0.8$ | 0.212 | 0.7 | 0.758 | 2.2 | 0.986 |
| $-2.2$ | 0.014 | $-0.7$ | 0.242 | 0.8 | 0.788 | 2.3 | 0.989 |
| $-2.1$ | 0.018 | $-0.6$ | 0.274 | 0.9 | 0.816 | 2.4 | 0.992 |
| $-2.0$ | 0.023 | $-0.5$ | 0.309 | 1.0 | 0.841 | 2.5 | 0.994 |
| $-1.9$ | 0.029 | $-0.4$ | 0.345 | 1.1 | 0.864 | 2.6 | 0.995 |
| $-1.8$ | 0.036 | $-0.3$ | 0.382 | 1.2 | 0.885 | 2.7 | 0.997 |
| $-1.7$ | 0.045 | $-0.2$ | 0.421 | 1.3 | 0.903 | 2.8 | 0.997 |
| $-1.6$ | 0.055 | $-0.1$ | 0.460 | 1.4 | 0.919 | 2.9 | 0.998 |
| $-1.5$ | 0.067 | 0.0 | 0.500 | 1.5 | 0.933 | 3.0 | 0.999 |

For $|x| \geqslant 3.3$, $N(x)$ may be taken as 0 or 1 as appropriate. More accurate tables can be found in standard compilations such as *Statistical tables for biological, agricultural and medical research* by R. A. Fisher and F. Yates (Longman, sixth edition reprinted 1982), but the values above are sufficient for present purposes.

come in whole numbers), we want the sum of the sample to be at least 39.5. But the distribution of this sum is approximately a normal distribution with mean $n/2 = 50$ and standard deviation $\sqrt{n/2} = 5$, so we can obtain an approximate answer by looking up Table 3.1 with $x = (39.5 - 50)/5$. This gives $x = -2.1$, whence $N(x) \approx 0.018$. Similarly, the probability that we get at most 60 heads (which means at most 60.5) is obtained by setting $x = (60.5 - 50)/5 = +2.1$, whence $N(x) \approx 0.982$. So the probability that we get between forty and sixty heads inclusive is approximately $(0.982 - 0.018)$, which rounds off to 0.96. We may not get fifty heads exactly, but we are unlikely to be very far away.

We can take this line of argument a little further. If we calculate the probability of obtaining between 47 and 53 heads inclusive, we find that it is approximately 0.52, so even this small target is more likely to be achieved than not. More generally, it is clear from Table 3.1 that $N(x)$ takes the values 0.25 and 0.75 at approximately $x = \pm 0.67$; in other words, the probability that a sample from a normal distribution lies within $\pm 0.67$ standard deviations from the mean is about 0.5. The standard deviation of $n$ tosses is $\sqrt{n}/2$, so we can expect to be within $\sqrt{n}/3$ heads of an exactly even result about half the time.[2]

# How random is a toss?

In the previous chapter, we questioned the randomness of practical card shuffling. Is the tossing of a coin likely to be any better?

The final state of a tossed coin depends on two things: its flight through the air, and its movement after landing. Let us look first at the flight. Consciously to bias the result of a toss by controlling the flight amounts to trying to predetermine the precise number of revolutions that the coin makes in the air, and this appears to be very difficult. I know of no experimental work on the subject, but a good toss should produce at least fifty revolutions of the coin, and it seems very unlikely that a perceptible bias can be introduced in such a toss. If the number of revolutions can be regarded as a sample from a normal distribution, even a bias as small as one part in ten thousand cannot be introduced unless the standard deviation is less than three quarters of a revolution. The true distribution is likely to differ slightly from normality and the bias may therefore be somewhat larger, but I still doubt if a bias exceeding one part in a thousand can be introduced in a spin of fifty revolutions.

The movement after landing is a much more likely cause of bias. Either an asymmetric mass or a bevelled edge may be expected to affect the result. The bias of a coin is unlikely to be large, but cowrie shells and cricket bats may well show a preference for a particular side.

But any bias that does exist can be greatly reduced by repetition. Suppose that the actual probability of a head is $(1+\epsilon)/2$ instead of $1/2$. If we toss twice, the probability of an even number of heads is now $(1+\epsilon^2)/2$; if we toss four times, it is $(1+\epsilon^4)/2$; and so on. Doubling the number of tosses and looking for an even number of heads squares the bias term $\epsilon$.

This is highly satisfactory. Let the probability of a head be 0.55, which is greater than anyone would assert of a normal coin; then the probability of an even number of heads from four tosses is only

---

[2] No, 0.67 doesn't represent two thirds; a more accurate calculation would revise it to 0.67449. But 'half a normal distribution lies within two thirds of a standard deviation from the mean' is a sufficient rule of thumb for many purposes.

0.50005. Even if the probability of a head is an utterly preposterous 0.9, we need only 42 tosses to bring the probability of an even number of heads below 0.50005.

We conclude that a well-tossed coin should be a very effective randomizer indeed.

# Cubic and other dice

We now turn to the other dice mentioned at the start of the chapter, and we start by considering the standard cubic die. If such a die is unbiased, each face has a probability of 1/6.

In itself, a cubic die is a less satisfactory randomizer than a coin. It certainly produces a greater range of results from a single throw, but the probabilities of these results are more likely to be unequal. Because it has more faces, the difficulties of accurate manufacture are greater; a classic statistical analysis by Weldon (1904) showed that his dice appeared to be biased, and a brief theoretical analysis by Roberts ('A theory of biased dice', *Eureka* **18** 8-11, 1955) showed that about a third of the bias observed in his own dice could be explained by the effect of asymmetric mass on the final fall. Roberts did not consider the effect of asymmetric mass on previous bouncing, and a complete analysis would undoubtedly be very difficult. Furthermore, the throw may easily introduce a bias. Any reasonably strong spin of an unbiased coin gives the two outcomes an acceptably even chance, but if a die is spun so that one face remains uppermost, or rolled so that two faces remain vertical, the outcomes are by no means of equal probability. Dice cheats have been known to take advantage of this.

In the last section, we saw that the bias of a coin can be reduced by tossing the same coin several times and looking for an even number of heads. A similar technique can be applied to a cubic die; we can throw the same die $m$ times, add the results, divide by six, and take the remainder. If the individual outcomes have probabilities bounded by $(1 \pm \epsilon)/6$ and the throw can be neglected as source of bias, it can be shown that the remainders have probabilities bounded by $(1 \pm \epsilon^m)/6$. But this procedure is not proof against malicious throwing, since a cheat with an interest in a particular remainder can bias his final throw accordingly.

Similar analyses can be applied to general polyhedral dice, to prismatic dice, and to teetotums and sectioned wheels. Dodecahedral and other polyhedral dice are even more difficult to manufacture accurately than cubic dice, and the problem of biased throwing remains. Prismatic dice are also difficult to manufacture accurately, and teetotums and sectioned wheels are more difficult still, because it is necessary not only to build an accurate polygon or wheel but to locate the axis of spin exactly at its centre; but prismatic dice, teetotums, and sectioned wheels all avoid the problem of biased throwing, since a reasonably strong roll or spin is as difficult to bias as a toss of a coin. Allegations of cheating at roulette always relate to the construction of the wheel or to deliberate interference as it slows down, not to the initial strength of the spin.

All this being said, the performance of any $n$-sided die can be improved by throwing it $m$ times, dividing the total by $n$, and taking the remainder. If the original outcomes have probabilities bounded by $(1 \pm \epsilon)/n$ and the throw can be neglected as source of bias, the remainders have probabilities bounded by $(1 \pm \epsilon^m)/n$.

# The arithmetic of dice games

In principle, the analysis of a dice game is just a matter of counting probabilities; the arithmetic may be tedious, but it is rarely difficult. We content ourselves with a few instructive cases. An unbiased cubic die is assumed throughout.

### (a) The winner of a race game

We start with a fundamental question. If the player due to throw next in a single-die race game is $y$ squares from the goal and his opponent is $z$ squares from it, what are his chances of winning?

When we considered a large number of tosses of a coin, we found that the Central Limit Theorem gave quick and accurate answers. It does so here as well. The relevant factors are as follows.

(i) The average distance moved by a throw is 7/2 squares, and the point midway between the players is now $(y+z)/2$ squares from the goal. So after a total of $2(y+z)/7$ throws, half by each player, we can expect this midpoint to be approximately at the goal. The probability that a player is ahead after this number of throws is therefore likely to be a reasonable approximation to the probability that he wins.

(ii) The standard deviation of a single throw is $\sqrt{(35/12)}$, so the spread of results after $n$ throws can be estimated by considering a normal distribution with standard deviation $\sqrt{(35n/12)}$. If $n = 2(y+z)/7$, this standard deviation becomes $\sqrt{\{5(y+z)/6\}}$.

(iii) The advantage of the throw is worth half the average distance moved, so the true advantage of the player due to throw next is $(z - y + 7/4)$.

So we want to find the probability that the player's opponent fails to gain $(z - y + 7/4)$ squares in $2(y+z)/7$ moves, and this can be estimated by setting $x$ to $(z - y + 7/4)/\sqrt{\{5(y+z)/6\}}$ and looking up $N(x)$ in Table 3.1. Detailed calculation shows that this formula yields an answer with an error of less than 0.01 provided that each player is at least ten squares from the goal.

And just as when tossing coins, we can draw conclusions which are perhaps surprising. If, being a hundred squares from the goal, you are a mere four squares behind and it is your opponent's throw, would you rate your chances as worse than two to one against? You should. Even the advantage of the throw itself is appreciable. If $x$ is small, $N(x)$ can be shown to be approximately $0.5 + x/\sqrt{(2\pi)}$, so if two players are the same distance $y$ from the goal, the probability that the player who is due to throw next will win is approximately $0.5 + \sqrt{(147/160\pi y)}$. Set $y = 25$, and this is greater than 0.6; set $y = 100$, and it is still greater than 0.55.

## (b) The probability of climbing a ladder

In practice, race games are usually spiced in some way: by forced detours, short cuts, bonuses, and penalties. The best known game of this type is 'snakes and ladders', in which a player who lands on the head of a snake slides down to its tail, and a player who lands at the foot of a ladder climbs up it. A simple question is now: given that there is a ladder $l$ squares ahead, what is the probability that we shall climb it? Our concern is only with the immediate future, and we ignore the possibility that the ground may be traversed again later in the game.

Table 3.2 gives the answer, both where there is no intervening snake (left-hand column) and where there is a single snake $s$ squares in front of the ladder. This table shows several interesting features. Even the peak at $l = 6$ in the first column may seem surprising at first sight, though a little reflection soon explains it: if we are six squares away and miss with our first throw, we must get another try. No other square guarantees two tries in this way.

**Table 3.2** Snakes and ladders: the probability of climbing a ladder

| $l$ | — | 1 | 2 | 3 | 4 | 5 | 6 | 7 | 8 | 9 | 10 |
|---|---|---|---|---|---|---|---|---|---|---|---|
| | | | | Square $s$ occupied by snake | | | | | | | |
| 1 | 0.17 | — | 0.17 | 0.17 | 0.17 | 0.17 | 0.17 | 0.17 | 0.17 | 0.17 | 0.17 |
| 2 | 0.19 | 0.17 | — | 0.19 | 0.19 | 0.19 | 0.19 | 0.19 | 0.19 | 0.19 | 0.19 |
| 3 | 0.23 | 0.19 | 0.19 | — | 0.23 | 0.23 | 0.23 | 0.23 | 0.23 | 0.23 | 0.23 |
| 4 | 0.26 | 0.23 | 0.23 | 0.23 | — | 0.26 | 0.26 | 0.26 | 0.26 | 0.26 | 0.26 |
| 5 | 0.31 | 0.26 | 0.26 | 0.26 | 0.26 | — | 0.31 | 0.31 | 0.31 | 0.31 | 0.31 |
| 6 | 0.36 | 0.31 | 0.31 | 0.31 | 0.31 | 0.31 | — | 0.36 | 0.36 | 0.36 | 0.36 |
| 7 | 0.25 | 0.19 | 0.19 | 0.19 | 0.19 | 0.19 | 0.19 | — | 0.25 | 0.25 | 0.25 |
| 8 | 0.27 | 0.23 | 0.20 | 0.20 | 0.20 | 0.20 | 0.20 | 0.23 | — | 0.27 | 0.27 |
| 9 | 0.28 | 0.24 | 0.23 | 0.20 | 0.20 | 0.20 | 0.20 | 0.23 | 0.24 | — | 0.28 |
| 10 | 0.29 | 0.24 | 0.24 | 0.23 | 0.19 | 0.19 | 0.19 | 0.23 | 0.24 | 0.24 | — |
| 12 | 0.29 | 0.24 | 0.23 | 0.23 | 0.22 | 0.21 | 0.16 | 0.21 | 0.22 | 0.23 | 0.23 |
| 14 | 0.28 | 0.24 | 0.23 | 0.22 | 0.21 | 0.20 | 0.19 | 0.22 | 0.19 | 0.20 | 0.21 |
| 16 | 0.29 | 0.24 | 0.23 | 0.22 | 0.21 | 0.20 | 0.18 | 0.22 | 0.22 | 0.22 | 0.18 |
| 18 | 0.29 | 0.24 | 0.23 | 0.22 | 0.21 | 0.20 | 0.18 | 0.21 | 0.21 | 0.21 | 0.21 |
| 20 | 0.29 | 0.24 | 0.23 | 0.22 | 0.21 | 0.20 | 0.18 | 0.21 | 0.21 | 0.20 | 0.20 |
| $\infty$ | 0.29 | 0.24 | 0.23 | 0.22 | 0.21 | 0.20 | 0.18 | 0.21 | 0.21 | 0.20 | 0.20 |

But it is the effect of the snakes that provides the greatest interest. It might seem that a snake immediately in front of the ladder ($s=1$) would provide the greatest obstacle, yet in fact it provides the least; a snake six squares in front is much more difficult to circumvent without missing the ladder as well. Even a very distant snake is a better guard than a snake within four squares of the ladder. Because the average distance moved by a throw is 7/2, the chance that we land on a particular square having come from a distance is approximately 2/7. So if we start at a sufficient distance beyond such a snake, the probability that we survive it is approximately 5/7, and the approximate probability that we hit the ladder in spite of it is therefore 10/49. This rounds to 0.20.

Indeed, if we come from a great distance, a single snake at $s=6$ is actually a better guard than a pair of snakes together at $s=1$ and $s=2$. In the absence of the snakes, our chance of hitting the ladder would be 2/7. In the presence of snakes at $s=1$ and $s=2$ (Figure 3.3), we can hit the ladder only by throwing a three or above, which reduces our chances by a third. This gives 4/21, which rounds to 0.19. A single snake at $s=6$ instead (Figure 3.4) reduces the probability of success to 0.18.

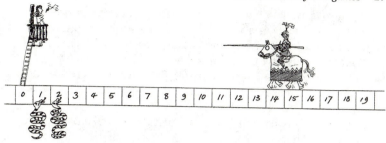

**Figure 3.3** Snakes and ladders: an apparently good guard

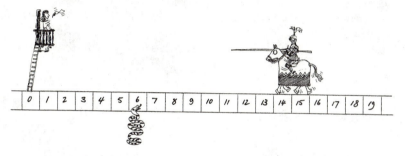

**Figure 3.4** Snakes and ladders: an even better guard

It is also curious that if we are within seven squares of the ladder and there is a single intervening snake, the probability of hitting the ladder is independent of the position of the snake. Yet a simple investigation shows why. The case $l=4$ is typical. In the absence of an intervening snake, the successful throws and combinations are 4, 3-1, 2-2, 1-3, 2-1-1, 1-2-1, 1-1-2, and 1-1-1-1. Every combination containing the same number of throws is equally likely; for example, each of the combinations 3-1, 2-2, and 1-3 has probability 1/36. A single snake, whatever its position, knocks out one two-throw combination, two three-throw combinations, and the four-throw combination; so each possible snake reduces the probability of success by the same amount.

## (c) Throwing an exact amount

Many games require a player to throw an exact amount. For example, a classic gambling swindle requires the punter to bet on the chance of obtaining a double six within a certain number of throws. The probability of obtaining a double six on a particular throw is 1/36, so it might seem that 18 throws would provide an even chance. Anybody who bets on this basis loses heavily. The probability of *not* getting a double six in 18 throws is $(35/36)^{18}$, which is greater than 0.6. Even with 24 throws, the odds are slightly against the thrower; 25 throws are needed before his chances exceed evens.

Even with only one die, a particular number may take a long time to materialize. Suppose that we need to throw a six, as the rules of family games frequently demand. More often than not, we succeed within four throws, but longer waits do happen. On one occasion in five, we must expect not to have succeeded within eight throws; on one occasion in ten, not within 12 throws; and on one occasion in a hundred, not even within 25 throws. This is an excellent recipe for childish tears, and parents may be well advised to modify the rules of games which make this demand.

# Simulation by computer

There is one form of die that we have not yet considered: the use of a sequence of computer-generated 'random' numbers. This is a specialized subject, but a brief mention is appropriate. Such a sequence can also be used to shuffle a card pack, and it is convenient to consider both topics together.

Computers behave very differently from humans. The success of conventional dice rests on the fact that humans cannot repeat complicated operations exactly. Computers can, and do. In fact the correct term for most computer-generated 'random' numbers is 'pseudo-random', since each number is obtained from its predecessors according to a fixed rule. The numbers are therefore not truly random, and the most for which we can hope is that their non-randomness be imperceptible in practice.

A typical computer generator is based on a formula which produces a sequence of integers each lying between 0 and $M-1$ inclusive, where $M$ is a large positive integer called the *modulus* of the generator. Each integer is then divided by $M$ to give a fraction lying between 0 and $0.999\ldots$, and this fraction is presented to the user as an allegedly random number. What happens next depends on the application. If we want to simulate an $n$-sided die, we multiply the fraction by $n$, take the integer part of the result, and add 1. This gives an integer between 1 and $n$ inclusive. If we want to shuffle a card pack, we lay out the pack in order, obtain a first number from the basic sequence, convert it to an integer $i_{52}$ between 1 and 52 inclusive, and interchange card $i_{52}$ with card 52; then we obtain a second number, convert it to an integer $i_{51}$ between 1 and 51 inclusive, and interchange card $i_{51}$ with card 51; and so on all the way down to $i_2$.

The effectiveness of the generator is therefore completely determined by the basic sequence of integers. In the simplest generators, this sequence is produced by what is known as the 'linear congruential' method: to obtain the next member, we multiply the current member by a large number $A$, add a second large number $C$, divide the result by $M$, and take the remainder. Provided that $A$, $C$, and $M$ are suitably chosen, it is possible to guarantee that every number from 0 to $M-1$ occurs in the long run with equal frequency, and this ensures that the resulting fractions are evenly distributed between 0 and 0.999 . . . (give or take the inevitable rounding errors).

However, straightforward linear congruential generators are just a little too simple to be used with confidence. There are always likely to be values of $k$ such that sets of $k$ successive numbers (or sets of numbers a distance $k$ apart) are undesirably correlated, and analysis for one value of $k$ throws very little light on the behaviour of the generator for other values. It is therefore better to use a generator in which two or more linear congruential sequences are combined. The simplest such generators are additive; the fractions from several generators are added, and the integral part of the result discarded. The constituent sequences should have different moduli, and no two moduli should share a common factor.

An even better technique, though it involves extra work, is to shuffle the numbers, using an algorithm due to MacLaren and Marsaglia. This algorithm involves two sequences: the sequence $x$ whose values are actually presented to the user (and which may itself be an additive combination of separate sequences), and a second sequence $s$ which is used only to reorder the first. Such a generator requires a buffer with space for $b$ numbers from the $x$ sequence, and the first $b$ numbers must be placed in it before use. The action of obtaining a number now involves four steps: obtaining the next number from the $s$ sequence, converting it into an integer between 1 and $b$ inclusive, presenting to the user the number currently occupying this position in the buffer, and refilling this position with the next number from the $x$ sequence. If possible, the buffer should hold several times as many numbers as are likely to be required at a time; when using the numbers to shuffle a card pack, for example, a buffer size of at least 256 is desirable. This algorithm is perhaps as good as can reasonably be expected; provided that the constituent sequences are sensibly chosen (in particular, that no two moduli share a common factor), it is very unlikely that the numbers will prove unsatisfactory.

Whatever its type, a generator usually requires the user to supply an initial 'seed' value. If only one sequence is required, the value of the seed is unlikely to matter, but if several sequences are required, each seed should be an independent random number (obtained, for example, by tossing a coin). If simply related numbers such as 1, 2, 3 are used as seeds, the resulting sequences may be undesirably correlated.

The generation of acceptably 'random' numbers by computer is a difficult task, far more difficult than the bland description in a typical home computer manual might suggest. This discussion has been inevitably brief, and readers who desire further information should consult *Seminumerical algorithms* by D. E. Knuth (Addison-Wesley, 1980) or Chapter 19 of my own *Practical computing for experimental scientists* (Oxford, 1988).

# 4

## TO ERR IS HUMAN

In the two previous chapters, we considered games whose play was governed entirely by chance. We now look at some games which would be free from chance effects if they were played perfectly. In practice, however, play is not perfect, and our purpose in this chapter is to see to what extent imperfections of play can be regarded as chance phenomena. We shall look only at golf, association football, and cricket, but similar analyses can be applied to many other games.

## Finding a hole in the ground

We start with golf, which is one of the simplest of all ball games. It is essentially a single-player game; a player has a ball and some clubs with which to strike it, and his object is to get the ball into a target hole using as few strokes as possible. Competitive play is achieved by comparing separate single-player scores.

Championship golf is usually played on a course with 18 holes. Each hole is surrounded by an area of very short grass (the 'green'); between the green and the starting point is a narrow area of fairly short grass (the 'fairway'); and on each side of the fairway is an area of long grass (the 'rough'). Also present may be sand traps ('bunkers'), bushes, trees, streams, lakes, spectators, and other obstacles. Four rounds over such a course make up a typical championship.

Now let us look at the effects of human error and uncertainty. It is convenient to start by considering play on the green. A modern championship green may be very much smoother than that portrayed in Figure 1.2, but the imperfection of the player's own action still introduces an element of uncertainty; from a typical distance, a few shots go straight in, most miss but finish so near to the hole that the next shot is virtually certain to succeed, and a few miss by so much that the next shot is likely to fail as well (Figure 4.1). A detailed analysis of muscle control might permit a distribution to be estimated,

**Figure 4.1** Golf: play near the hole

but we shall not proceed to that level. Our point is that a player who has just arrived on the green can expect sometimes to take one more stroke, usually to take two, and occasionally to take three or even more.

The play to the green is subject to a similar but greater scatter. If the green is within range, the player's objective is to finish on it. There is a definite probability that he does so, and a smaller probability that he even finishes sufficiently near to the hole to need only one stroke on the green instead of the two that are more usually required. There is even a very small probability that he goes straight into the hole from off the green. On the other hand, he may fail to finish on the green, in which case several things may happen: he may finish on the fairway, probably still within range of the green and perhaps sufficiently near to the hole to have a good chance of getting down in only two more strokes; he may finish in the rough or behind an obstacle, perhaps in a position from which he will do well even to reach the green with his next shot (for the lie of a ball in the rough or among obstacles is very much a matter of luck even on a championship course); or he may finish in an unplayable position, perhaps in a lake or outside the boundaries of the course, in which case the rules compel him to accept a penalty and drop a new ball back at the position from which he has just played. This is an ignominy which occurs occasionally even in championship golf, and rather more frequently in the works of P. G. Wodehouse.

It would be very difficult to analyse the cumulative effect of these chance factors directly, so the most promising approach is to obtain some actual scores and calculate their standard deviation. But to obtain a suitable set of scores is not quite as easy as it might seem. The spread of a player's scores depends on his expertise, being greater for a novice than for a champion, and the rules of competitions are designed to select a winner and not to shed light on mathematical theories. So the best that can be done is to examine the scores of the leading players in a typical major championship, and to regard their spread as a measure of the *minimum* spread that can be expected in practice.

The 1985 British Open Championship provides a suitable example.[1] This championship consisted of four rounds, and was contested by 150 players. All played the first two rounds, those who scored 149 or better over two rounds continued into the third round, and those who scored 221 or better over three rounds played out the last round. This produced a total of 60 four-round scores, which are reproduced in Table 4.1.

It may now seem that we can estimate the effect of chance factors on a player's score simply by averaging the standard deviation of each set of four scores, but there are several complications. The first, which is easily accommodated, is that the most appropriate average is not the mean but the root mean square. The second is that conditions during the four rounds were not the same; in particular, the weather on the second day was abnormally bad, and the average score in the second round was appreciably higher than those in other rounds. Any such differences will have introduced additional variation, and the easiest way of allowing for them is to subtract the average score for the round from each individual score before calculating each player's standard deviation. The conditions under which such a simple correction is valid are discussed in statistical textbooks; suffice it to say that the errors introduced by its use here appear to be negligible.

The third complication is that the standard deviation of a sample of $n$ units does not correctly estimate the standard deviation of the set from which it is drawn; it underestimates it by a factor $\sqrt{\{(n-1)/n\}}$. A proof is outside the scope of this book, but a simple example is instructive. Suppose that we have a set of numbers consisting of ones and zeros in equal probability: heads and tails, if you like. We saw in the previous chapter that the standard deviation of such a set is $1/2$. Now suppose that we estimate this standard deviation by drawing a sample of two units. Half the time, the sample contains a one and a zero, and its standard deviation is indeed $1/2$; but the rest of the time, the sample contains two equal numbers, and its standard deviation is 0. If we perform this operation repeatedly and form the root mean square of the resulting standard deviations, we obtain $1/2\sqrt{2}$, which is an underestimate by a factor of $\sqrt{(1/2)}$ as predicted by the formula. Now what we have in Table 4.1 is a sample of four scores

---

[1] There is no particular significance in this choice, nor in that of any other example in this chapter. I simply used the data that were most conveniently to hand.

**Table 4.1** Golf: analysis of scores in the British Open Championship, 1985

| Rounds 1 | 2 | 3 | 4 | Total | $s_{60}$ | $s_{38}$ | Rounds 1 | 2 | 3 | 4 | Total | $s_{60}$ | $s_{38}$ |
|---|---|---|---|---|---|---|---|---|---|---|---|---|---|
| 68 | 71 | 73 | 70 | 282 | 1.9 | 2.3 | 71 | 73 | 72 | 73 | 289 | 0.3 | 1.0 |
| 70 | 75 | 70 | 68 | 283 | 2.6 | 2.0 | 68 | 73 | 74 | 74 | 289 | 2.3 | 2.9 |
| 74 | 72 | 70 | 68 | 284 | 3.1 | 2.6 | 71 | 78 | 70 | 71 | 290 | 3.1 | 2.6 |
| 64 | 76 | 72 | 72 | 284 | 4.2 | 4.4 | 71 | 70 | 77 | 72 | 290 | 3.5 | 3.9 |
| 70 | 72 | 70 | 72 | 284 | 0.5 | 0.9 | 75 | 74 | 71 | 71 | 291 | 2.4 | 1.9 |
| 72 | 69 | 68 | 75 | 284 | 3.3 | 3.7 | 73 | 75 | 70 | 73 | 291 | 1.7 | 1.3 |
| 68 | 71 | 70 | 75 | 284 | 2.4 | 3.0 | 72 | 72 | 73 | 74 | 291 | 1.2 | 1.8 |
| 70 | 76 | 69 | 70 | 285 | 2.6 | 2.1 | 75 | 73 | 68 | 75 | 291 | 3.4 | 3.3 |
| 69 | 71 | 74 | 71 | 285 | 2.1 | 2.5 | 68 | 72 | 80 | 72 | 292 | 5.0 | — |
| 73 | 73 | 67 | 72 | 285 | 2.8 | 2.5 | 71 | 73 | 76 | 72 | 292 | 2.2 | — |
| 76 | 68 | 74 | 68 | 286 | 5.0 | 4.9 | 75 | 74 | 70 | 73 | 292 | 2.3 | — |
| 75 | 72 | 70 | 69 | 286 | 3.2 | 2.8 | 73 | 74 | 72 | 73 | 292 | 0.7 | — |
| 72 | 75 | 70 | 69 | 286 | 2.5 | 1.8 | 70 | 74 | 72 | 76 | 292 | 1.9 | — |
| 69 | 76 | 70 | 71 | 286 | 2.4 | 2.0 | 74 | 73 | 71 | 75 | 293 | 1.8 | — |
| 71 | 74 | 68 | 73 | 286 | 2.1 | 2.0 | 71 | 70 | 80 | 72 | 293 | 4.9 | — |
| 71 | 75 | 72 | 69 | 287 | 2.3 | 1.7 | 74 | 72 | 71 | 76 | 293 | 2.4 | — |
| 74 | 74 | 69 | 70 | 287 | 2.8 | 2.2 | 72 | 74 | 75 | 73 | 294 | 1.3 | — |
| 71 | 72 | 71 | 73 | 287 | 0.7 | 1.3 | 72 | 75 | 74 | 73 | 294 | 0.8 | — |
| 70 | 71 | 71 | 75 | 287 | 2.0 | 2.6 | 69 | 74 | 77 | 74 | 294 | 3.0 | — |
| 71 | 73 | 74 | 70 | 288 | 2.0 | 1.9 | 73 | 76 | 71 | 74 | 294 | 1.6 | — |
| 72 | 72 | 73 | 71 | 288 | 1.4 | 1.5 | 73 | 75 | 71 | 75 | 294 | 1.5 | — |
| 73 | 76 | 68 | 71 | 288 | 3.1 | 2.5 | 70 | 71 | 76 | 77 | 294 | 3.5 | — |
| 71 | 78 | 66 | 73 | 288 | 4.4 | 4.0 | 72 | 73 | 72 | 77 | 294 | 2.1 | — |
| 71 | 75 | 69 | 73 | 288 | 2.0 | 1.7 | 73 | 73 | 75 | 74 | 295 | 1.4 | — |
| 68 | 76 | 77 | 68 | 289 | 4.7 | 4.6 | 71 | 70 | 77 | 77 | 295 | 3.9 | — |
| 70 | 76 | 73 | 70 | 289 | 2.4 | 2.1 | 70 | 75 | 75 | 77 | 297 | 2.4 | — |
| 72 | 76 | 71 | 70 | 289 | 2.3 | 1.7 | 70 | 79 | 70 | 78 | 297 | 4.1 | — |
| 70 | 74 | 73 | 72 | 289 | 1.1 | 1.3 | 69 | 74 | 75 | 80 | 298 | 4.0 | — |
| 73 | 72 | 72 | 72 | 289 | 1.3 | 1.4 | 74 | 71 | 74 | 81 | 300 | 4.3 | — |
| 72 | 74 | 71 | 72 | 289 | 0.9 | 0.4 | 76 | 72 | 73 | 80 | 301 | 3.8 | — |

| Average standard deviation (root mean square) | 2.8 | 2.6 |
|---|---|---|

Columns $s_{60}$ and $s_{38}$ measure the standard deviation of each player's score, adjusted to take account of column means calculated from the first 60 and 38 scores respectively. For details of the calculation, see the text.

from each player, so we must multiply each player's standard deviation by $\sqrt{(4/3)}$ to obtain a true estimate.

If we perform this calculation for each of the scores in Table 4.1, subtracting the average score for the round from each individual score, calculating the standard deviation of each player's adjusted scores, and multiplying by $\sqrt{(4/3)}$, we obtain the values in column $s_{60}$ of Table 4.1. The value at the foot of this column is the root mean square of the individual values.

We now come to the fourth and most difficult complication. The imposition of cuts at 149 after two rounds and 221 after three means that high scores in the early rounds have been eliminated from the data, whereas high scores in the last round remain. This means that the last-round mean is disproportionately high. There is no fully satisfactory way of dealing with this complication, but examination shows that there appears to be a threshold between 291 and 292; all but two of the players whose totals exceeded 291 might have been eliminated before the last round had they made their scores in a different order, whereas only four of the players whose totals did not exceed 291 might have been. Since our objective is to estimate a minimum spread by examining the scores of the leading players, it therefore seems best to restrict our analysis to the 38 players whose scores add up to no more than 291. This is done in column $s_{38}$ of Table 4.1. Our conclusion is that the effect of chance factors *on this particular course* causes a champion golfer's score over 18 holes to show a standard deviation of at least 2.6; and since the standard deviation of a score increases with the square root of the number of holes played, his scores over 72 holes can be expected to show twice this variation.

Another way of looking at this figure is instructive. Each hole on a golf course has a 'par' score which a first-class player should normally achieve. Over 72 holes, a standard deviation of 5.2 is equivalent to scoring exact par on two holes out of every three, and being within one stroke of par on all but one of the rest. Such a performance is indeed typical of first-class golfers.

Now let us suppose for a moment that the distribution of a player's 72-hole scores is approximately normal. (In truth, it is almost certain to be slightly skewed, since it is easier for bad luck to waste strokes than for good luck to gain them, but any appreciable deviation from normality is likely to increase still further the variability that we are about to see.) Half a normal distribution is more than 0.67 standard deviations away from the mean, and over a tenth of it is more than

1.6 standard deviations away. So over half the time, a leading golfer can expect his score for 72 holes to be more than three strokes away from his theoretical expectation, *even if his intrinsic performance remains constant;* and over a tenth of the time, he can expect it to be more than eight strokes away. For what it is worth, the 1984 champion scored 292 in 1985, and finished well down the list.

# Finding a hole in the defence

We now turn to association football. This is a ball game played between teams, each of which defends a goal and attempts to send the ball through the opponent's goal. Play is for a fixed period, and the team scoring the larger number of goals during this time is the winner. If the scores are equal, the game is drawn.

The role played by chance in this game is easily seen. In a typical situation near goal, an attacking player has only a moment to control a moving ball and play it so that it evades the final defender (the 'goalkeeper') and enters the goal. In stories for boys, the hero succeeds every time; in real life, he may play the ball within reach of the goalkeeper, or he may beat the goalkeeper only to miss the goal as well. Figure 4.2 shows the various possibilities in diagrammatic form. If each such attempt has a probability $p$ of success, the effect is the same as we obtain by tossing a biased coin; we have a binomial distribution. In particular, if a team makes twenty attempts to score and each has a probability 0.1 of success, the probability that it scores precisely $r$ goals is shown in the first column of Table 4.2. It will be noted that the resulting variability is quite high; for example, there is a probability of slightly greater than one in eight that it scores four

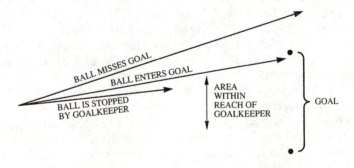

**Figure 4.2** Association football: play near the goal

**Table 4.2** A comparison of binomial and Poisson distributions

| r | $B_{20}$ | P | r | $B_{20}$ | P |
|---|---|---|---|---|---|
| 0 | 0.122 | 0.135 | 5 | 0.032 | 0.036 |
| 1 | 0.270 | 0.271 | 6 | 0.009 | 0.012 |
| 2 | 0.285 | 0.271 | 7 | 0.002 | 0.003 |
| 3 | 0.190 | 0.180 | 8 | 0.000 | 0.001 |
| 4 | 0.090 | 0.090 | 9 | 0.000 | 0.000 |

Column $B_{20}$ shows the probability of $r$ successes, assuming 20 attempts each with probability 0.1; column $P$ shows the same probability assuming a Poisson distribution with mean 2.

goals or more, and almost as high a probability that it does not score at all. So the effect of chance factors near the goal appears to provide at least a partial explanation for the observed variability of football scores.

To go further, we must examine some data. Sets of actual results are readily available, and in some respects they are very suitable for analysis; in particular, the all-play-all fixture list of a typical league produces a reasonably balanced set of results, even though changes in personnel mean that teams are not quite of constant strength throughout a season. But there are two difficulties: published results merely record the numbers of goals scored and not the numbers of scoring attempts, and not all attempts have an equal chance of success. It is therefore not clear how many separate binomial distributions should be assumed, nor what the associated probabilities should be.

However, goals are relatively rare events; a typical first-class team scores fewer than two per game. Random rare events obey the Poisson distribution, which states that if an event occurs on average $m$ times in a time interval, then the probability that it occurs precisely $r$ times in a particular interval is $e^{-m}m^r/r!$ where $e = 2.718\ldots$. We may therefore be able to replace an unknown combination of binomial distributions by a single Poisson distribution without committing too large an error. The Poisson distribution, with $m=2$, is shown in the second column of Table 4.2, and it can be seen that its agreement with the first column is quite close.[2]

[2] It is the Poisson distribution which justifies the remark, made in Chapter 2, that if an event has a very low probability $p$ then it is necessary to perform approximately $0.7/p$ trials to obtain an even chance of its occurrence. Let $n$ be the number of trials required; then the average number of occurrences within this number of trials is $np$. The probability that there will be no occurrence in a particular set of $n$ trials is therefore $e^{-np}$, and this reaches 1/2 when $n$ reaches $0.7/p$ approximately.

The next task is to estimate a suitable expected score *m*. The earliest investigations of football scores (for example, that by Moroney in *Facts from figures*, Penguin, 1956) assumed the same expected score for all teams in all matches. The very limited computing power then available made this a necessary simplification, but it is plainly unrealistic and modern computers permit a more detailed approach. In fact, three major factors determine a team's expectation in a particular game: its own strength, the strength of its opponents, and the ground on which the game is being played (because a team playing on its own ground has a distinct advantage). In a typical league, every team plays every other team twice, once on each ground, and this permits these factors to be estimated with reasonable accuracy; for example, the total number of goals scored by a team is a reasonable measure of its strength in attack. So in a game between teams *i* and *j*, played on *i*'s ground, we might as a first approximation take *i*'s expected score to be $F_i A_j H / N^2$, where $F_i$ is the total number of goals scored by team *i* against all other teams, $A_j$ is the total number of goals conceded by team *j* against all other teams, *H* is the total number of goals scored by teams playing on their own ground, and *N* the total number scored by all teams. Similarly, we might take the expectation of team *j* in the same match to be $F_j A_i (N - H) / N^2$. These formulae do not represent any deep mathematical insight; they are merely the simplest formulae which allow for the advantage of the home team and ensure that if every team were to score exactly to expectation in every match then the total numbers of goals scored and conceded by every team would be as actually happened.

Table 4.3 shows an analysis on this basis of the 1983-4 results from the four divisions of the English Football League (known at that time as the Canon League, in consequence of sponsorship). Using our assumptions, it is a simple matter to calculate the expected score of each team in each of these games and then to calculate the probability of every possible score. Each set of three values in the table shows the predicted number of occurrences of a score, the actual number of occurrences, and the difference.

Table 4.3 shows that our predictions are at least moderately sensible; we do not predict 50 occurrences of a result when the actual number is 25 or 100. In view of the crudity of our assumptions, this is really rather gratifying. A close examination nevertheless shows some weaknesses. We have overestimated the number of teams failing to score, both at home and away, and we have underestimated scores of two or three goals by home teams and one goal by away teams. We

**Table 4.3** Association football: analysis of results in the Canon League, 1983-4

| Away score | 0 | 1 | 2 | Home score 3 | 4 | 5 | 6+ |
|---|---|---|---|---|---|---|---|
| 0 | 153 | 214 | 160 | 103 | 45 | 28 | 10 |
|   | 160.9 | 235.0 | 187.6 | 108.4 | 50.6 | 20.3 | 10.6 |
|   | −7.9 | −21.0 | −27.6 | −5.4 | −5.6 | 7.7 | −0.6 |
| 1 | 149 | 244 | 203 | 98 | 41 | 16 | 8 |
|   | 157.9 | 221.5 | 170.2 | 94.8 | 42.7 | 16.6 | 8.2 |
|   | −8.9 | 22.5 | 32.8 | 3.2 | −1.7 | −0.6 | −0.2 |
| 2 | 61 | 98 | 104 | 62 | 17 | 12 | 3 |
|   | 84.6 | 114.0 | 84.4 | 45.4 | 19.8 | 7.4 | 3.5 |
|   | −23.6 | −16.0 | 19.6 | 16.6 | −2.8 | 4.6 | −0.5 |
| 3 | 32 | 34 | 43 | 24 | 5 | 4 | 2 |
|   | 32.8 | 42.4 | 30.3 | 15.8 | 6.7 | 2.4 | 1.1 |
|   | −0.8 | −8.4 | 12.7 | 8.2 | −1.7 | 1.6 | 0.9 |
| 4+ | 10 | 19 | 14 | 10 | 2 | 0 | 0 |
|   | 13.9 | 17.1 | 11.6 | 5.8 | 2.3 | 0.7 | 0.4 |
|   | −3.9 | 1.9 | 2.4 | 4.2 | −0.3 | −0.7 | −0.4 |

Each pair of values shows the actual number $A$ of occurrences of the result and the number $P$ predicted by Poisson distributions as described in the text. The subtraction shows the discrepancy $A - P$.

have also overestimated the number of away wins and underestimated the number of draws. That these are not just chance effects is shown by Table 4.4, which shows the differences between actual and predicted occurrences for each of the four divisions individually, classified by home score, away score, and margin of result. Another revealing comparison appears in Table 4.5, which shows the difference between the actual and predicted numbers of points scored by each team over the season (the rather curious scoring system being 3 for a win, 1 for a draw, and 0 for a loss). The predictions for the first division are remarkably good, but the others show a general tendency to underestimate strong teams and overestimate weak ones.

So we have some systematic discrepancies, and it is easy to suggest possible causes.

(a) Our crude formula for a team's expected score against a particular opponent is unlikely to be completely accurate.

**Table 4.4** Association football: further analysis of results in the Canon League, 1983–4

| Score or margin | | Div. 1 A | A−P | Div. 2 A | A−P | Div. 3 A | A−P | Div. 4 A | A−P | Total A | A−P |
|---|---|---|---|---|---|---|---|---|---|---|---|
| Home score | 0 | 107 | −5.1 | 102 | −7.9 | 98 | −9.3 | 98 | −22.8 | 405 | −45.1 |
| | 1 | 135 | −10.8 | 136 | −8.6 | 154 | −12.6 | 184 | 11.1 | 609 | −21.0 |
| | 2 | 112 | 5.2 | 114 | 7.5 | 155 | 17.2 | 143 | 9.9 | 524 | 39.8 |
| | 3 | 70 | 12.3 | 70 | 11.7 | 81 | 0.3 | 76 | 2.5 | 297 | 26.8 |
| | 4 | 22 | −3.5 | 18 | −8.4 | 41 | 3.5 | 29 | −3.8 | 110 | −12.3 |
| | 5 | 10 | 0.3 | 18 | 7.5 | 16 | 1.2 | 16 | 3.4 | 60 | 12.4 |
| | 6+ | 6 | 1.4 | 4 | −1.7 | 7 | −0.3 | 6 | −0.2 | 23 | −0.8 |
| Away score | 0 | 160 | −12.3 | 167 | −17.1 | 192 | −12.8 | 194 | −18.2 | 713 | −60.5 |
| | 1 | 178 | 17.7 | 170 | 10.1 | 196 | −0.5 | 215 | 19.9 | 759 | 47.1 |
| | 2 | 71 | −12.7 | 86 | 7.9 | 113 | 12.5 | 87 | −9.8 | 357 | −2.2 |
| | 3 | 34 | 1.9 | 27 | −1.4 | 41 | 4.6 | 42 | 7.4 | 144 | 12.5 |
| | 4+ | 19 | 5.5 | 12 | 0.5 | 10 | −3.7 | 14 | 0.8 | 55 | 3.0 |
| Home win by | 5+ | 10 | 2.5 | 13 | 3.3 | 11 | −0.3 | 12 | 1.4 | 46 | 6.9 |
| | 4 | 13 | −1.4 | 15 | −1.2 | 20 | −1.0 | 16 | −3.2 | 64 | −6.7 |
| | 3 | 37 | 3.7 | 31 | −4.7 | 43 | −4.0 | 47 | 3.4 | 158 | −1.7 |
| | 2 | 56 | −9.5 | 61 | −7.0 | 96 | 8.3 | 66 | −17.7 | 279 | −25.9 |
| | 1 | 110 | 7.9 | 114 | 10.8 | 116 | −11.0 | 144 | 18.3 | 484 | 26.0 |
| Draw | | 118 | 5.0 | 117 | 5.9 | 142 | 13.5 | 150 | 18.0 | 527 | 42.4 |
| Away win by | 1 | 68 | −6.1 | 66 | −4.5 | 91 | 11.5 | 73 | −10.0 | 298 | −9.1 |
| | 2 | 29 | −5.4 | 34 | 2.3 | 22 | −12.4 | 23 | −13.6 | 108 | −29.1 |
| | 3 | 13 | 0.4 | 10 | −1.4 | 8 | −3.5 | 17 | 4.4 | 48 | −0.1 |
| | 4+ | 8 | 2.9 | 1 | −3.7 | 3 | −1.1 | 4 | −0.8 | 16 | −2.8 |

Each pair of values shows the actual number $A$ of occurrences, and the difference $A−P$ between this number and the number $P$ predicted by Poisson distributions.

(b) No allowance has been made for changes in the pattern of play after a goal has been scored.

(c) The membership of teams is not constant; players may miss games through injury, or be transferred from one team to another during the season.

(d) The probability of a particular score has been estimated by a single Poisson distribution instead of by the set of binomial distributions which we believe really to exist.

The last of these possibilities is the easiest to examine. A Poisson distribution provides quite a good approximation to a binomial distribution, as we saw in Table 4.2, but it underestimates the frequencies of scores near to the expected mean and overestimates the frequencies of very low scores, and this is precisely the effect that we

**Table 4.5** Association football: analysis of points scored in the Canon League, 1983-4

| Team | Div. 1 | | Div. 2 | | Div. 3 | | Div. 4 | |
|---|---|---|---|---|---|---|---|---|
| | A | A−P | A | A−P | A | A−P | A | A−P |
| 1 | 80 | −4.6 | 88 | 1.0 | 95 | 7.1 | 101 | 3.6 |
| 2 | 77 | 0.8 | 88 | 5.7 | 87 | 11.4 | 85 | 4.5 |
| 3 | 74 | −3.1 | 80 | 3.5 | 83 | −2.3 | 82 | 1.7 |
| 4 | 74 | −3.1 | 70 | 0.8 | 83 | −0.4 | 82 | 1.8 |
| 5 | 73 | −4.6 | 70 | 3.8 | 79 | 6.7 | 75 | 7.1 |
| 6 | 63 | −3.5 | 67 | 2.1 | 75 | 7.1 | 72 | −2.7 |
| 7 | 62 | 3.8 | 64 | 1.9 | 71 | 2.4 | 68 | −10.6 |
| 8 | 61 | 3.5 | 61 | 6.0 | 70 | 3.1 | 67 | −6.5 |
| 9 | 60 | −0.9 | 60 | −3.7 | 67 | −0.4 | 67 | 6.9 |
| 10 | 60 | 3.4 | 60 | 2.7 | 64 | 3.4 | 66 | 3.3 |
| 11 | 57 | 3.6 | 57 | −5.4 | 63 | 4.4 | 63 | −0.4 |
| 12 | 53 | −3.4 | 57 | −5.3 | 62 | −10.8 | 62 | 1.2 |
| 13 | 52 | 2.6 | 57 | 5.6 | 62 | 8.5 | 60 | −1.9 |
| 14 | 51 | −5.5 | 52 | −8.4 | 61 | −0.4 | 59 | −2.7 |
| 15 | 51 | −5.4 | 51 | 0.6 | 61 | 5.3 | 59 | −2.5 |
| 16 | 51 | 1.2 | 49 | −14.7 | 56 | −7.7 | 59 | 2.6 |
| 17 | 51 | 2.4 | 49 | −3.8 | 55 | −2.8 | 58 | −6.3 |
| 18 | 50 | 4.9 | 47 | −3.5 | 54 | −4.9 | 53 | 4.5 |
| 19 | 50 | 3.4 | 47 | 4.1 | 51 | −8.4 | 52 | −9.3 |
| 20 | 48 | −0.9 | 42 | 6.3 | 49 | −9.4 | 48 | −6.6 |
| 21 | 41 | −3.6 | 29 | −1.4 | 46 | −5.8 | 48 | 3.4 |
| 22 | 29 | 4.0 | 24 | −3.8 | 44 | −7.0 | 46 | −0.8 |
| 23 | | | | | 43 | −1.8 | 40 | −1.1 |
| 24 | | | | | 33 | −10.7 | 34 | −7.1 |

Each pair of values shows the actual number $A$ of points, and the difference $A-P$ between this number and the number $P$ predicted by Poisson distributions.

see in Tables 4.3 and 4.4. This suggests that we should replace the Poisson distribution with a set of $n$ binomial distributions with probabilities $p_1, \ldots, p_n$ where $p_1 + \ldots + p_n = m$. But published results give no data from which suitable values of $n$ and $p_1, \ldots, p_n$ can be estimated. Reep and Benjamin, in an extensive analysis of games played between 1953 and 1967 ('Skill and chance in association football', *Journal of the Royal Statistical Society A* **131** (1968), 581-5) found an almost constant ratio 10 between shots and goals, so $n$ might be taken as $10m$, but even this does not tell us what values should be assumed for $p_1, \ldots, p_n$. It is hardly realistic that they should be assumed equal, yet if one or two are relatively high, and the rest very low, the standard deviation of the distribution is sharply reduced. It is certainly possible to choose a set of binomial distributions that would give a better fit to the actual results than our Poisson distribution, but there is little point in doing this as long as the other problems remain unresolved.

All this being said, however, the agreement of our simple model with reality is distinctly encouraging, and it demonstrates that the chance effects inherent in goal-scoring appear to go a very long way towards explaining the observed variation of football results.[3]

# A game of glorious uncertainty

For our last example, we consider the archetypally English game of cricket.

Cricket is a team game, but at its heart is an individual contest between a batsman and a bowler. The bowler sends a ball at a target (the 'wicket') which is defended by the batsman. If the ball hits the wicket, the batsman is 'out', and another batsman takes his place; and even if the batsman prevents the ball from hitting the wicket, he may still be out in other ways (most usually, by hitting the ball in the air so that a 'fielder' catches it before it bounces). If the batsman avoids all these fates, he may score 'runs' in various ways, and the team whose batsmen score the greater number of runs is the winner.

From a mathematical point of view, the dominant factor is the probability that an error by the batsman causes him to be out. Superficially, this is akin to the probability that a scoring attempt succeeds at football; for example, if the bowler is trying to beat the batsman by making the ball swerve, he may fail to make it swerve sufficiently, or he may beat the batsman only to miss the wicket as

---

[3] Readers familiar with the standard $\chi^2$ test of practical statistics may wonder why we have not applied it. If this test is applied to Table 4.3, it shows the discrepancies to be about half as large again as would have been expected had they been due solely to chance, and even as modest an excess as this is highly significant; a table based on as many results as Table 4.3 can be expected to show such an excess only once every thousand trials. But this merely tells us that our simple model is almost certainly not quite good enough (which doesn't surprise us in the slightest, since we know that our assumptions are crude), whereas the patterns displayed in Tables 4.4 and 4.5 throw light on what its deficiencies are likely to be. The assessment of differences between theoretical predictions and actual data depends both on their magnitude *and on their pattern;* and while computers may be better at calculating magnitude, it is pattern that is usually the more revealing.

well.[4] However, there is a fundamental difference, in that a game of football stops after a certain time, whereas a batsman's innings at cricket usually continues until he is out. This causes the mathematical nature of the game to be quite different.

To investigate it, let us assume that every time a ball is bowled there is a probability $p$ that the batsman is out. In this case, there is a probability $p$ that he is out on his first ball, a probability $p(1-p)$ that he survives his first ball only to be out on his second, and a probability $p(1-p)^{n-1}$ that he is out on his $n$th. It can now be shown that the average number of balls that he survives before being out is $(1-p)/p$. Furthermore, if he scores an average of $r$ runs from each ball that he survives, his average score is $r(1-p)/p$. In first-class play, typical values for a good batsman are $r=0.4$ and $p=0.01$; in other words, he expects to score at a rate of around two runs every five balls, and to survive about a hundred balls before being out.

Now let us consider the way in which a batsman's scores may be expected to be distributed around his average score. Suppose that this average score is $m$. In order to score $km$ runs, he must survive $km/r$ balls, and the probability that he does so is $(1-p)^{km/r}$. But $m=r(1-p)/p$, so this probability reduces to $(1-p)^{k(1-p)/p}$; and for all realistically small values of $p$, this can be shown to be approximately equal to $e^{-k}$, where $e=2.718\ldots$ as usual.[5] It follows that a first-class batsman whose average score is $m$ may be expected to score fewer than $m/10$ on about one occasion in ten, fewer than $m/2$ on about two occasions in five, and fewer than $m$ on about nineteen occasions in thirty; yet he may be expected to score more than $2m$ on about four occasions in thirty, and more than $3m$ on about one occasion in twenty. This indicates the range of performance that may be expected from a particular batsman *even if the underlying probabilities are constant from one occasion to the next*.

Another instructive way of looking at this distribution is to consider its standard deviation. This can be shown to be approximately equal

---

[4] Although the attack on the wicket is fundamental to cricket, a first-class batsman is most frequently out in practice because he hits the ball in the air and a fielder catches it, and the bowler often tries not so much to hit the wicket as to deceive the batsman into a mishit from which a catch can be made. This makes the role of chance even clearer. Sometimes the bowler fails to deceive the batsman at all; sometimes he deceives him so much that he misses the ball altogether; sometimes the batsman does indeed mishit, but not within reach of a fielder; and sometimes the fielder drops the ball.

[5] A standard theorem of algebra states that the product $(1-k/n)^n$ tends to $e^{-k}$ as $n$ increases, so $(1-p)^{k/p}$ tends to $e^{-k}$ as $p$ tends to 0. The effect of the factor $(1-p)$ in the exponent is negligible.

to the mean $m$, in contrast to the Poisson and binomial distributions, where the standard deviation is approximately equal to the square root of the mean (exactly $\sqrt{m}$ in the case of the Poisson, and $\sqrt{\{(1-p)m\}}$ in the case of the binomial), and to the observed distribution of golf scores, where it is only a small fraction of the square root of the mean. It follows that golf scores cluster relatively closely, football scores vary substantially, and cricket scores vary enormously.

A consequence is that we need a large amount of data to estimate a batsman's average score with precision. The Central Limit Theorem, which we used in Chapter 3 to estimate the behaviour of sums, also applies to means, and indeed it is usually stated in textbooks in this form; it asserts that if we repeatedly sample a population with mean $m$ and standard deviation $s$, the mean of the sample approaches a normal distribution with the same mean $m$ but with standard deviation $s/\sqrt{n}$ where $n$ is the number of items in the sample. So if we have a typical first-class batsman whose true average is 40 and we compute his actual average over 25 innings, this computed average has a standard deviation of 8 *even if conditions are constant from one game to the next*. Yet batting averages are regularly published to two decimal places, and they are lapped up by readers as if this precision were meaningful.

It is time to look at some real data. This proves to be unexpectedly difficult. Vast amounts of numerical data exist, but the conditions under which they have been gathered are so variable that it is difficult to draw other than the crudest of conclusions. The standard of nominally 'first-class' teams varies from weak university sides to international touring teams full of fearsomely fast bowlers. The most extensive English first-class competition, the County Championship, does not always provide balanced results, since some teams may play each other twice whereas others meet only once; and in any case, the players comprising a team may differ widely from one match to another (much more widely than at football, because international matches are played in parallel with county matches and teams may be depleted by international commitments). A team normally uses several bowlers during an innings. Weather and ground conditions may affect the batsman and the bowler unequally, and the tactical state of the game may compel either the batsman or the bowler to play abnormally. All these things affect a batsman's expectation in any particular game. There is also a purely technical complication, which we have ignored until now, in that a batsman may not be out at all. This does not affect his average expectation, if we redefine it as the ratio between the number of runs he scores over a period and the number of times that he is out, but it certainly affects the number of runs that he is likely to score on any particular occasion.

All this being said, however, the County Championship provides as good a set of data as can reasonably be obtained, and the batsmen's scores for 1980 (when sponsorship caused it to be known as the Schweppes County Championship) are summarized in Table 4.6. The contents of this table allow for the various complications inherent in

**Table 4.6** Cricket: analysis of batsmen's scores in the Schweppes County Championship, 1980

| | | | | | Notional expected score | | | | | | |
|---|---|---|---|---|---|---|---|---|---|---|---|
| | 15 | | 25 | | 35 | | 45 | | 55 | | 65 |
| k | N | p | N | p | N | p | N | p | N | p | N | p |
| 0.2 | 616 | 0.24 | 960 | 0.23 | 1028 | 0.24 | 387 | 0.24 | 166 | 0.22 | 116 | 0.18 |
| 0.4 | 514 | 0.15 | 795 | 0.16 | 845 | 0.16 | 322 | 0.15 | 132 | 0.15 | 85 | 0.25 |
| 0.6 | 410 | 0.17 | 653 | 0.16 | 658 | 0.20 | 251 | 0.20 | 99 | 0.23 | 68 | 0.18 |
| 0.8 | 327 | 0.17 | 532 | 0.17 | 547 | 0.16 | 203 | 0.18 | 77 | 0.21 | 52 | 0.20 |
| 1.0 | 276 | 0.13 | 426 | 0.17 | 446 | 0.17 | 171 | 0.13 | 62 | 0.14 | 44 | 0.14 |
| 1.2 | 231 | 0.15 | 341 | 0.16 | 360 | 0.16 | 140 | 0.15 | 50 | 0.15 | 36 | 0.18 |
| 1.4 | 190 | 0.14 | 267 | 0.17 | 292 | 0.16 | 106 | 0.20 | 42 | 0.14 | 32 | 0.06 |
| 1.6 | 156 | 0.15 | 222 | 0.13 | 231 | 0.17 | 82 | 0.20 | 32 | 0.20 | 18 | 0.42 |
| 1.8 | 135 | 0.12 | 184 | 0.13 | 182 | 0.18 | 67 | 0.13 | 27 | 0.16 | 14 | 0.18 |
| 2.0 | 109 | 0.13 | 150 | 0.16 | 154 | 0.12 | 54 | 0.16 | 20 | 0.09 | 12 | 0.14 |
| 2.4 | 81 | 0.22 | 100 | 0.29 | 96 | 0.31 | 25 | 0.49 | 11 | 0.31 | 10 | 0.09 |
| 2.8 | 56 | 0.28 | 67 | 0.28 | 60 | 0.33 | 17 | 0.23 | 3 | 0.6 | 8 | 0.1 |
| 3.2 | 40 | 0.22 | 47 | 0.24 | 38 | 0.30 | 9 | 0.18 | 1 | 0.5 | 5 | 0.0 |
| 3.6 | 32 | 0.18 | 26 | 0.41 | 20 | 0.35 | 3 | 0.5 | 0 | 1.0 | 1 | 0.5 |
| 4.0 | 20 | 0.33 | 19 | 0.21 | 11 | 0.31 | 1 | 0.5 | 0 | — | 0 | 1.0 |
| 4.4 | 15 | 0.17 | 7 | 0.46 | 9 | 0.18 | 1 | 0.0 | 0 | — | 0 | — |
| 4.8 | 10 | 0.23 | 1 | 0.8 | 7 | 0.1 | 0 | — | 0 | — | 0 | — |
| 5.2 | 3 | 0.6 | 1 | 0.0 | 2 | 0.6 | 0 | — | 0 | — | 0 | — |
| 5.6 | 1 | 0.5 | 1 | 0.0 | 1 | 0.5 | 0 | — | 0 | — | 0 | — |
| 6.0 | 1 | 0.0 | 0 | 1.0 | 0 | 1.0 | 0 | — | 0 | — | 0 | — |

The batsmen have been classified into decades according to their average scores over the season, and the midpoint of each decade has been taken as a notional expected score for all batsmen within it. Each row represents a fraction $k$ of this expected score, and each pair of values shows (i) the number of times $N$ a batsman reached this fraction of his expected score, and (ii) the proportion $p$ of batsmen who reached the previous fraction but were then out having failed to reach this fraction. Batsmen who were left not out between two fractions have been ignored when calculating the corresponding proportion.

cricket, in particular for the fact that a batsman's innings may finish other than by his being out. It excludes batsmen who were out fewer than ten times or whose average score was less than 10, and groups the rest by average into the six ranges 10–19.99 . . . , 20–29.99 . . . , and so on. The midpoint of each range is then taken as a mean expectation for the column, and the values in the left-hand column indicate fractions of this expectation. The tabular values show (i) the number of times that a batsman reached this fraction, and (ii) the proportion of batsmen who reached the previous fraction but were then out before reaching this fraction, batsmen who were left not out between the two fractions being ignored when calculating this proportion. If the probabilities underlying a batsman's expectation are indeed constant, these proportions should be approximately constant also (except that the doubled vertical scale from row 2.4 onwards means that the proportions in these rows are comparable with approximately twice the proportions in the earlier rows.) In fact the agreement across the rows is generally good, particularly if we discount proportions which are based on only a few scores, but a systematic decrease is observable down the columns, which is what we would expect if a batsman's expectation does change from occasion to occasion. Furthermore, the high proportions in row 0.2 bear out the widely held belief that a batsman is most vulnerable early in his innings.

Because of the widely differing conditions under which cricket is played, Table 4.6 is much the least satisfactory of the tables in this chapter. Nevertheless, it is clear that a large proportion of the observed variability of cricket scores can be explained by the effect of chance on the dismissal of a batsman. It is customary for the more hysterical of sporting journalists to praise a batsman to the skies when he scores two consecutive hundreds, and to condemn him utterly when he suffers two consecutive failures. Praise for hundreds is fair enough, since skill as well as luck is needed, but exaggerated condemnation of failures merely betrays an ignorance of the laws of mathematics.

# 5

## IF A BEATS B, AND B BEATS C . . .

In the previous chapter, we looked at some of the pseudo-random effects which appear to affect the results of games. We now attempt to measure the actual skill of performers. There is no difficulty in finding apparently suitable mathematical formulae; textbooks are full of them. Our primary aim here is to examine the circumstances in which a particular formula may be valid, and to note any difficulties which may attend its use.

## The assessment of a single player in isolation

We start by considering games such as golf, in which each player records an independent score. In practice, of course, few competitive games are completely free from interactions between the players; a golfer believing himself to be two strokes behind the tournament leader may take risks that he would not take if he believed himself to be two strokes ahead of the field. But for present purposes, we assume that any such interactions can be ignored. We also ignore any effects that external circumstances may have on our data. In Chapter 4, we were able to adjust our scores to allow for the general conditions pertaining to each round, because the pooling of the scores of all the players allowed the effect of these conditions to be assessed with reasonable confidence. A sequence of scores from one player alone does not allow such assessments to be made, and we have little alternative but to accept the scores at face value.

To fix our ideas, let us suppose that a player has returned four separate scores, say 73, 71, 70, and 68 (Figure 5.1). If these scores were recorded at approximately the same time, we might conclude that a reasonable estimate of his skill is given by the unweighted mean 70.5 (*U* in Figure 5.1). This is effectively the basis on which tournament results are calculated. On the other hand, if the scores were returned

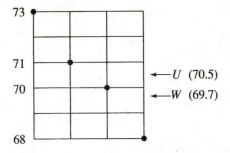

**Figure 5.1** Weighted and unweighted means

over a long period, we might prefer to give greater weight to the more recent of them. For example, if we assign weights 1:2:3:4 in order, we obtain a weighted mean of 69.7 ($W$ in Figure 5.1). More sophisticated weighting, taking account of the actual dates of the scores, is also possible.

So we see, right from the start, that our primary need is not a knowledge of abstruse formulae, but a commonsense understanding of the circumstances in which the data have been generated.

Now let us assume that we already have an estimate, and that the player returns an additional score. Specifically, let us suppose that our estimate has been based on $n$ scores $s_1, \ldots, s_n$, and that the player has now returned an additional score $s_{n+1}$. If we are using an unweighted mean based on the $n$ most recent scores, we must now replace our previous estimate

$$(s_1 + \ldots + s_n)/n$$

by a new estimate

$$(s_2 + \ldots + s_{n+1})/n;$$

the contribution from $s_1$ vanishes, the contributions from $s_2, \ldots, s_n$ remain unchanged, and a new contribution appears from $s_{n+1}$. In other words, the contribution of a particular score to an unweighted mean remains constant until $n$ more scores have been recorded, and then suddenly vanishes. On the other hand, if we use a weighted mean with weights 1:2: . . . :$n$, the effect of a new score $s_{n+1}$ is to replace the previous estimate

$$2(s_1 + 2s_2 + \ldots + ns_n)/n(n+1)$$

by a new estimate

$$2(s_2 + 2s_3 + \ldots + ns_{n+1})/n(n+1);$$

not only does the contribution from $s_1$ vanish, but the contributions from $s_2, \ldots, s_n$ are all decreased. This seems rather more satisfactory.

Nevertheless, anomalies may still arise. Let us go back to the scores in Figure 5.1, which yielded a mean of 69.7 using weights 1:2:3:4, and let us suppose that an additional score of 70 is recorded. If we form a new estimate by discarding the earliest score and applying the same weights 1:2:3:4 to the remainder, we obtain 69.5, which is less than either the previous estimate or the additional score. So we check our arithmetic, suspecting a mistake, but we find the value indeed to be correct. Such an anomaly is always possible when the mean of the previous scores differs from the mean of the contributions discarded. It is rarely large, but it may be disconcerting to the inexperienced.

If we are to avoid anomalies of this kind, we must ensure that the updated estimate always lies between the previous estimate and the additional score. This is easily done; if $E_n$ is the estimate after $n$ scores $s_1, \ldots, s_n$, all we need is to ensure that

$$E_{n+1} = w_n E_n + (1 - w_n)s_{n+1}$$

where $w_n$ is some number satisfying $0 < w_n < 1$. But there is a cost. If we calculate successive estimates $E_1, E_2, \ldots$, we find

$$E_1 = s_1,$$

$$E_2 = w_1 s_1 + (1 - w_1)s_2,$$

$$E_3 = w_1 w_2 s_1 + w_2(1 - w_1)s_2 + (1 - w_2)s_3,$$

and so on; the contribution of each score gradually decreases, but it never vanishes altogether.

So we have a fundamental dilemma. If we want to ensure that an updated estimate always lies between the most recent score and the previous estimate, we must accept that even the most ancient of scores will continue to contribute its mite to our estimate. Conversely, if we exclude scores of greater than a certain antiquity, we must be prepared for occasions on which an updated estimate does not lie between the previous estimate and the most recent score.

# The estimation of trends

The estimates that we have discussed so far have assessed skill as it has been displayed in the past. If a player's skill has changed appreciably during the period under assessment, the estimate may not fully reflect the change. It is therefore natural to try to find estimates which do reflect such changes.

Such estimates can indeed be made. Figure 5.2(a) repeats the last two data values of Figure 5.1, and the dotted line shows the estimate $E$ obtained by assuming that the change from 70 to 68 represents a genuine trend that may be expected to continue. More sophisticated estimates, taking account of more data values, can be found in textbooks on statistics and economics.

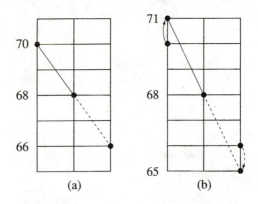

**Figure 5.2** The behaviour of a forward estimate

But there are two problems. The first, which is largely a matter of common sense, is that the assumption of a trend is not to be made blindly. Golf enthusiasts may have recognized 73-71-70-68 as the sequence returned by Ben Hogan when winning the British Open in 1953, and it is doubtful if even Hogan, given a fifth round, would have gone round a course as difficult as Carnoustie in 66. On the other hand, there are circumstances in which the same figures might much more plausibly indicate a trend: if they represent the successive times of a twenty-kilometre runner as he gets into training, for example.

The second difficulty is a matter of mathematics. The extrapolation from 70 through 68 to 66 is an example of a linear extrapolation from $s_1$ through $s_2$, the estimate being given by $2s_2 - s_1$. In other words, we form a weighted mean of $s_1$ and $s_2$, but one of the weights is negative. A change in that score therefore has an inverse effect on the estimate. This is shown in Figure 5.2(b), where the score 70 has been changed to 71 and it is seen that the estimate has changed from 66 to 65. In particular, if a negatively weighted score was abnormally poor (for example, because the player was not fully fit on that occasion), the future estimate will be *improved* as a result.

This contravenes common sense, and suggests that we should confine our attention to estimates which respond conformably to all constituent scores: a decrease in any score should decrease the estimate, and an increase in any score should increase it. But it turns out that such an estimate cannot lie outside the bounds of the constituent scores, and this greatly reduces the scope for estimation of trends. The proof is simple and elegant. Let $S$ be the largest of the constituent scores. If each score actually equals $S$, the estimate must equal $S$ also. If any score $s$ does not equal $S$ and the estimating procedure is conformable, the replacement of $S$ by $s$ must cause a reduction in the estimate. So a conformable estimate cannot exceed the largest of the constituent scores; and similarly, it cannot be less than the smallest of them.[1]

In practice, therefore, we have little choice. Given that common sense demands conformable behaviour, we cannot use an estimating procedure which predicts a future score outside the bounds of previous scores; we can merely give the greatest weight to the most recent of them. If this is unwelcome news to improving youngsters, it is likely to gratify old stagers who do not like being reminded too forcibly of their declining prowess. In fact, the case which most commonly causes difficulty is that of a player who has recently entered top-class competition and whose first season's performance is appreciably below the standard which he subsequently establishes; and the best way to handle this case is not to use a clever formula to estimate the improvement, but to ignore the first year's results when calculating subsequent estimates.

# Interactive games

We now turn to games in which the result is recorded only as a win for a particular player, or perhaps as a draw. These games present a much more difficult problem. The procedure usually adopted is to assume that the performance of a player can be represented by a single number, called his *grade* or *rating*, and to calculate this grade so as to reflect his actual results. For anything other than a trivial

---

[1] It follows that economic estimates which attempt to project current trends are in general *not* conformable; and while this is unlikely to be the whole reason for their apparent unreliability, it is not an encouraging thought.

game, the assumption is a gross over-simplification, so anomalies are almost inevitable and controversy must be expected. In the case of chess, which is the game for which grading has been most widely adopted, a certain amount of controversy has indeed arisen; some players and commentators appear to regard grades with excessive reverence, most assume them to be tolerable approximations to the truth, a few question the detailed basis of calculation, and a few regard them as a complete waste of ink. The resolution of such controversy is beyond the scope of this book, but at least we can illuminate the issues.

The basic computational procedure is to assume that the mean expected result of a game between two players is given by an 'expectation function' which depends only on their grades $a$ and $b$, and then to calculate these grades so as to reflect the actual results. It might seem that the accuracy of the expectation function is crucial, but we shall see in due course that it is actually among the least of our worries; provided that the function is reasonably sensible, the errors introduced by its inaccuracy are likely to be small compared with those resulting from other sources. In particular, if the game offers no advantage to either player, it may well be sufficient to calculate the grading difference $d=a-b$ and to use a simple smooth function $f(d)$ such as that shown in Figure 5.3. For a game such as chess, the function should be offset to allow for the first player's advantage, but this is a detail easily accommodated.[2]

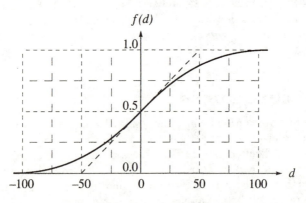

**Figure 5.3** A typical expectation function

[2] Figure 5.3 adopts the chess player's scaling of results: 1 for a win, 0 for a loss, and 0.5 for a draw. The scaling of the $d$-axis is arbitrary.

Once the function $f(d)$ has been chosen, the calculation of grades is straightforward. Suppose for a moment that two players already have grades which differ by $d$, and that they now play another game, the player with the higher grade winning. Before the game, we assessed his expectation as $f(d)$; after the game, we might reasonably assess it as a weighted mean of the previous expectation and the new result. Since a win has value 1, this suggests that his new expectation should be given by a formula such as

$$w + (1 - w)f(d)$$

where $w$ is a weighting factor, and this is equivalent to

$$f(d) + w\{1 - f(d)\}.$$

More generally, if the stronger player achieves a result of value $r$, the same argument suggests that his new expectation should be given by the formula

$$f(d) + w\{r - f(d)\}.$$

Now if the expectation function is scaled as in Figure 5.3 and the grading difference $d$ is small, we see that a change of $\delta$ in $d$ produces a change of approximately $\delta/100$ in $f(d)$. It follows that approximately the required change in expectation can be obtained by increasing the grading difference by $100w\{r - f(d)\}$. As the grading difference becomes larger, the curve flattens, and a given change in the grading difference produces a smaller change in the expectation. In principle, this can be accommodated by increasing the scaling factor 100, but it is probably better to keep this factor constant, since always to make the same change in the expectation may demand excessive changes in the grades. The worst case occurs when a player unexpectedly fails to beat a much weaker opponent; the change in grading difference needed to reduce an expectation of 0.99 to 0.985 may be great indeed. To look at the matter another way, keeping the scaling factor constant amounts to giving reduced weight to games between opponents of widely differing ability, which is plainly reasonable since the ease with which a player beats a much weaker opponent does not necessarily say a great deal about his ability against his approximate peers.

A simple modification of this procedure can be used to assign a grade to a previously ungraded player. Once he has played a reasonable number of games, he can be assigned that grade which would be left unchanged if adjusted according to his actual results. The same technique can also be used if it is desired to ignore ancient history and grade a player only on the basis of recent games.

Grades calculated on this basis can be expected to provide at least a rough overall measure of each regular player's performance. However, certain practical matters must be decided by the grading administrator, and these may have a perceptible effect on the figures. Examples are the interval at which grades are updated, the value of the weighting parameter $w$, the relative division of an update between the grades of the players (in particular, when one player is well established whereas the other is a relative newcomer), the criteria by which less than fully competitive games are excluded, and the circumstances in which a player's grade is recalculated to take account only of his most recent games. Grades are therefore not quite the objective measures that their more uncritical admirers like to maintain.

# Grades as measures of ability

Although grading practitioners usually stress that their grades are merely measures of *performance*, players are interested in them primarily as measures of *ability*. A grading system defines an expectation between every pair of graded players, and the grades are of interest only in so far as these expectations correspond to reality.

A little thought suggests that this correspondence is unlikely to be exact. If two players $A$ and $B$ have the same grade, their expectations against any third player $C$ are asserted to be exactly equal. Alternatively, suppose that $A$, $B$, $Y$, and $Z$ have grades such that $A$'s expectation against $B$ is asserted to equal $Y$'s against $Z$, and that expectations are calculated using a function which depends only on the grading difference. If these grades are $a$, $b$, $y$, and $z$, then they must satisfy $a-b=y-z$, from which it follows that $a-y=b-z$, and hence $A$'s expectation against $Y$ is asserted to equal $B$'s against $Z$. Assertions as precise as this are unlikely to be true for other than very simple games, and it follows that grades cannot be expected to yield exact expectations; the most for which we can hope is that they form a reasonable average measure whose deficiencies are small compared with the effects of chance fluctuation.

These chance effects can easily be estimated. If $A$'s expectation against $B$ is $p$ and there is a probability $h$ that they draw, the standard deviation $s$ of a single result is $\sqrt{(\{p(1-p)-h/4\})}$. If they now play a sufficiently long series of $n$ games, the distribution of the discrepancy between mean result and expectation can be taken as a normal distribution with standard deviation $s/\sqrt{n}$, and a simple rule of thumb gives the approximate probability that any particular discrepancy would have arisen by chance: a discrepancy exceeding the standard deviation can be expected on about one trial in three, and a discrepancy exceeding twice the standard deviation on about one trial in twenty. What constitutes a sufficiently large value of $n$ depends on the expectation $p$. If $p$ lies between 0.4 and 0.6, $n$ should be at least 10; if $p$ is smaller than 0.4 or greater than 0.6, $n$ should be at least $4/p$ or $4/(1-p)$ respectively. More detailed calculations, taking into account the incidence of each specific combination of results, are obviously possible, but they are unlikely to be worthwhile.

A practicable testing procedure now suggests itself. Every time a new set of grades is calculated, the results used to calculate the new grades can be used also to test the old ones. If two particular opponents play each other sufficiently often, their results provide a particularly convenient test; otherwise, results must be grouped, though this must be done with care since the grouping of inhomogeneous results may lead to wrong conclusions. The mean of the new results can be compared with the expectation predicted by the previous grades, and large discrepancies can be highlighted: one star if the discrepancy exceeds the standard deviation, and two if it exceeds twice the standard deviation. The rule of thumb above gives the approximate frequency with which stars are to be expected if chance fluctuations are the sole source of error.

In practice, of course, chance fluctuations are not the only source of error. Players improve when they are young, they decline as they approach old age, and they sometimes suffer temporary loss of form due to illness or domestic disturbance. The interpretation of stars therefore demands common sense. Nevertheless, if the proportions of stars and double stars greatly exceed those attributable to chance fluctuation, the usefulness of the grades is clearly limited.

If grades do indeed constitute acceptable measures of ability, regular testing such as this should satisfy all but the most extreme and blinkered of critics. However, grading administrator and critic alike must always remember that *around one discrepancy in three should be starred, and around one in twenty doubly starred, on account of chance fluctuations, even if there is no other source of error.* If a grading administrator performs a hundred tests without finding any doubly starred discrepancies, he should not congratulate himself on the success of his grading system; he should check the correctness of his testing.

# The self-fulfilling nature of grading systems

We now come to one of the most interesting mathematical aspects of grading systems: their self-fulfilling nature. It might seem that a satisfactory expectation function must closely reflect the true nature of the game, but in fact this is not so. Regarded as measures of ability, grades are subject to errors from two sources: (i) discrepancies between ability and actual performance, and (ii) errors in the calculated expectations due to the use of an incorrect expectation function. In practice, the latter are likely to be much smaller than the former.

Table 5.1 illustrates this. It relates to a very simple game in which each player throws a single object at a target, scoring a win if he hits and his opponent misses, and the game being drawn if both hit or if both miss. If the probability that player $j$ hits is $p_j$, the expectation of player $j$ against player $k$ can be shown to be $(1 + p_j - p_k)/2$, so we can calculate expectations exactly by setting the grade of player $j$ to $50p_j$ and using the expectation function $f(d) = 0.5 + d/100$. Now let us suppose that we have nine players whose probabilities $p_1, \ldots, p_9$ range linearly from 0.1 to 0.9, that they play each other with equal frequency, and that we deliberately use the incorrect expectation function $f(d) = N(d\sqrt{(2\pi)}/100)$ where $N(x)$ is the normal distribution function. The first column of Table 5.1 shows the grades that are produced if the results of the games agree strictly with expectation, and the entries for each pair of players show (i) the discrepancy between the true and the calculated expectations, and (ii) the standard deviation of a single result between the players. The latter is always large compared with the former, which means that a large number of games are needed before the discrepancy can be detected against the background of chance fluctuations. The standard deviation of a mean result decreases only with the inverse square root of the number of games played, so we can expect to require well over a hundred sets of all-play-all results before even the worst discrepancy (player 1 against player 9) can be diagnosed with confidence.

Experiment bears this out. Table 5.2 records a computer simulation of a hundred sets of all-play-all results, the four rows for each player showing (i) his true expectation against each opponent, (ii) the mean of his actual results against each opponent, (iii) his grade as calculated from these results using the correct expectation function $0.5 + d/100$, together with his expectation against each opponent as calculated from their respective grades, and (iv) the same as calculated using the

**Table 5.1** Throwing one object: the effect of an incorrect expectation function

| Player | Grade | Opponent | | | | | | | | |
|---|---|---|---|---|---|---|---|---|---|---|
| | | 1 | 2 | 3 | 4 | 5 | 6 | 7 | 8 | 9 |
| 1 | 5.5 | — | −0.009 | −0.013 | −0.014 | −0.011 | −0.005 | 0.004 | 0.017 | 0.032 |
| | | — | 0.250 | 0.274 | 0.287 | 0.292 | 0.287 | 0.274 | 0.250 | 0.212 |
| 2 | 17.3 | 0.009 | — | −0.006 | −0.009 | −0.009 | −0.007 | −0.002 | 0.006 | 0.017 |
| | | 0.250 | — | 0.304 | 0.316 | 0.320 | 0.316 | 0.304 | 0.283 | 0.250 |
| 3 | 28.5 | 0.013 | 0.006 | — | −0.004 | −0.006 | −0.007 | −0.005 | −0.002 | 0.004 |
| | | 0.274 | 0.304 | — | 0.335 | 0.339 | 0.335 | 0.324 | 0.304 | 0.274 |
| 4 | 39.3 | 0.014 | 0.009 | 0.004 | — | −0.003 | −0.006 | −0.007 | −0.007 | −0.005 |
| | | 0.287 | 0.316 | 0.335 | — | 0.350 | 0.346 | 0.335 | 0.316 | 0.287 |
| 5 | 50.0 | 0.011 | 0.009 | 0.006 | 0.003 | — | −0.003 | −0.006 | −0.009 | −0.011 |
| | | 0.292 | 0.320 | 0.339 | 0.350 | — | 0.350 | 0.339 | 0.320 | 0.292 |
| 6 | 60.7 | 0.005 | 0.007 | 0.007 | 0.006 | 0.003 | — | −0.004 | −0.009 | −0.014 |
| | | 0.287 | 0.316 | 0.335 | 0.346 | 0.350 | — | 0.335 | 0.316 | 0.287 |
| 7 | 71.5 | −0.004 | 0.002 | 0.005 | 0.007 | 0.006 | 0.004 | — | −0.006 | −0.013 |
| | | 0.274 | 0.304 | 0.324 | 0.335 | 0.339 | 0.335 | — | 0.304 | 0.274 |
| 8 | 82.7 | −0.017 | −0.006 | 0.002 | 0.007 | 0.009 | 0.009 | 0.006 | — | −0.009 |
| | | 0.250 | 0.283 | 0.304 | 0.316 | 0.320 | 0.316 | 0.304 | — | 0.250 |
| 9 | 94.5 | −0.032 | −0.017 | −0.004 | 0.005 | 0.011 | 0.014 | 0.013 | 0.009 | — |
| | | 0.212 | 0.250 | 0.274 | 0.287 | 0.292 | 0.287 | 0.274 | 0.250 | — |

The grades are calculated using an incorrect expectation function as described in the text. The tabular values show (i) the discrepancy between the calculated and true expectations, and (ii) the standard deviation of a single result.

incorrect expectation function $N(d\sqrt{(2\pi)}/100)$. The differences between rows (i) and (iii) are caused by the differences between the theoretical expectations and the actual results, and the differences between rows (iii) and (iv) are caused by the difference between the expectation functions. In over half the cases, the former difference is greater than the latter, so on this occasion even a hundred sets of all-play-all results have not sufficed to betray the incorrect expectation function with reasonable certainty. Nor are the differences between actual results and theoretical expectations in Table 5.2 in any way abnormal. If the experiment were to be performed again, it is slightly more likely than not that the results in row (ii) would differ from expectation more widely than those which appear here.[3]

[3] In practice, of course, we do not know the true expectation function, so rows (i) and (iii) are hidden from us, and all we can do is to assess whether the discrepancies between rows (ii) and (iv) might reasonably be attributable to chance. Such a test is far from sensitive; for example, the discrepancies in Table 5.2 are so close to the median value which can be expected from chance fluctuations alone that nothing untoward can be discerned in them. We omit a proof of this, because the analysis is not straightforward; the simple rules of thumb which we used in the previous section cannot

**Table 5.2** Throwing one object: grading systems compared

| Player | Grade | Opponent | | | | | | | | |
|---|---|---|---|---|---|---|---|---|---|---|
| | | 1 | 2 | 3 | 4 | 5 | 6 | 7 | 8 | 9 |
| 1 | | — | 0.450 | 0.400 | 0.350 | 0.300 | 0.250 | 0.200 | 0.150 | 0.100 |
| | | — | 0.455 | 0.435 | 0.350 | 0.335 | 0.230 | 0.200 | 0.150 | 0.125 |
| | 11.8 | — | 0.471 | 0.400 | 0.355 | 0.314 | 0.250 | 0.197 | 0.182 | 0.110 |
| | 7.8 | — | 0.466 | 0.388 | 0.342 | 0.304 | 0.247 | 0.203 | 0.191 | 0.139 |
| 2 | | 0.550 | — | 0.450 | 0.400 | 0.350 | 0.300 | 0.250 | 0.200 | 0.150 |
| | | 0.545 | — | 0.395 | 0.395 | 0.330 | 0.290 | 0.245 | 0.210 | 0.130 |
| | 17.6 | 0.529 | — | 0.429 | 0.384 | 0.344 | 0.280 | 0.226 | 0.211 | 0.139 |
| | 14.6 | 0.534 | — | 0.422 | 0.374 | 0.334 | 0.275 | 0.228 | 0.215 | 0.159 |
| 3 | | 0.600 | 0.550 | — | 0.450 | 0.400 | 0.350 | 0.300 | 0.250 | 0.200 |
| | | 0.565 | 0.605 | — | 0.450 | 0.390 | 0.380 | 0.315 | 0.285 | 0.185 |
| | 31.7 | 0.600 | 0.570 | — | 0.455 | 0.414 | 0.350 | 0.297 | 0.281 | 0.209 |
| | 30.4 | 0.612 | 0.578 | — | 0.451 | 0.409 | 0.344 | 0.292 | 0.277 | 0.212 |
| 4 | | 0.650 | 0.600 | 0.550 | — | 0.450 | 0.400 | 0.350 | 0.300 | 0.250 |
| | | 0.650 | 0.605 | 0.550 | — | 0.435 | 0.430 | 0.365 | 0.310 | 0.240 |
| | 40.8 | 0.645 | 0.616 | 0.546 | — | 0.459 | 0.395 | 0.343 | 0.327 | 0.254 |
| | 40.2 | 0.658 | 0.626 | 0.549 | — | 0.457 | 0.390 | 0.336 | 0.320 | 0.249 |
| 5 | | 0.700 | 0.650 | 0.600 | 0.550 | — | 0.450 | 0.400 | 0.350 | 0.300 |
| | | 0.665 | 0.670 | 0.610 | 0.565 | — | 0.370 | 0.395 | 0.370 | 0.305 |
| | 48.9 | 0.685 | 0.657 | 0.586 | 0.540 | — | 0.436 | 0.383 | 0.368 | 0.295 |
| | 48.8 | 0.696 | 0.666 | 0.591 | 0.543 | — | 0.432 | 0.376 | 0.360 | 0.284 |
| 6 | | 0.750 | 0.700 | 0.650 | 0.600 | 0.550 | — | 0.450 | 0.400 | 0.350 |
| | | 0.770 | 0.710 | 0.620 | 0.570 | 0.630 | — | 0.395 | 0.435 | 0.395 |
| | 61.7 | 0.750 | 0.721 | 0.650 | 0.604 | 0.564 | — | 0.447 | 0.432 | 0.359 |
| | 62.4 | 0.753 | 0.725 | 0.656 | 0.610 | 0.568 | — | 0.442 | 0.425 | 0.345 |
| 7 | | 0.800 | 0.750 | 0.700 | 0.650 | 0.600 | 0.550 | — | 0.450 | 0.400 |
| | | 0.800 | 0.755 | 0.685 | 0.635 | 0.605 | 0.605 | — | 0.520 | 0.400 |
| | 72.3 | 0.803 | 0.773 | 0.703 | 0.685 | 0.617 | 0.553 | — | 0.484 | 0.412 |
| | 74.0 | 0.797 | 0.772 | 0.708 | 0.664 | 0.624 | 0.558 | — | 0.483 | 0.400 |
| 8 | | 0.850 | 0.800 | 0.750 | 0.700 | 0.650 | 0.600 | 0.550 | — | 0.450 |
| | | 0.850 | 0.790 | 0.715 | 0.690 | 0.630 | 0.565 | 0.480 | — | 0.425 |
| | 75.4 | 0.818 | 0.789 | 0.718 | 0.673 | 0.633 | 0.569 | 0.516 | — | 0.427 |
| | 77.5 | 0.809 | 0.785 | 0.723 | 0.680 | 0.640 | 0.575 | 0.517 | — | 0.417 |
| 9 | | 0.900 | 0.850 | 0.800 | 0.750 | 0.700 | 0.650 | 0.600 | 0.550 | — |
| | | 0.875 | 0.870 | 0.815 | 0.760 | 0.695 | 0.605 | 0.600 | 0.575 | — |
| | 89.9 | 0.891 | 0.861 | 0.791 | 0.745 | 0.705 | 0.641 | 0.588 | 0.573 | — |
| | 94.3 | 0.861 | 0.841 | 0.788 | 0.751 | 0.716 | 0.655 | 0.600 | 0.583 | — |

For each player, the four rows show (i) the true expectation against each opponent; (ii) the average result of a hundred games against each opponent, simulated by computer; (iii) the grade calculated from the simulated games, using the correct expectation function, and the resulting expectations against each opponent; and (iv) the same using an incorrect expectation function as described in the text.

be applied, because we are now looking at the spread of results around expectations *to whose calculation they themselves have contributed* (whereas the rules apply to the spread of results about *independently calculated* expectations) and we must take the dependence into account. Techniques exist for doing this, but the details are beyond the scope of this book.

This is excellent news for grading secretaries, since it suggests that any reasonable expectation function can be used; the spacing of the grades may differ from that which a correct expectation function would have generated, but the expectations will be adjusted in approximate compensation, and any residual errors will be small compared with the effect of chance fluctuation on the actual results. But there is an obvious corollary: *the apparently successful calculation of expectations by a grading system throws no real light on the underlying nature of the game.* Chess grades are currently calculated using a system, due to A. E. Elo, in which expectations are calculated by the normal distribution function, and the general acceptance of this system by chess players has fostered the belief that the normal distribution provides the most appropriate expectation function for chess. In fact it is by no means obvious that this is so. The normal distribution function is not a magic formula of universal applicability; its validity as an estimator of unknown chance effects depends on the Central Limit Theorem, which states that the *sum* of a *large number* of *independent* samples from the same distribution can be regarded as a sample from a normal distribution, and it can reasonably be adopted as a model for the behaviour of a game only if the chance factors affecting a result are equivalent to a large number of independent events which combine additively. Chess may well not satisfy this condition, since many a game appears to be decided not by an accumulation of small blunders but by a few large ones. But while the question is of some theoretical interest, it hardly matters from the viewpoint of practical grading. Chess gradings are of greatest interest at master level, and the great majority of games at this level are played within an expectation range of 0.3 to 0.7. Over this range, the normal distribution function is almost linear, but so is any simple alternative candidate, and so in all probability is the unknown 'true' function which most closely approximates to the actual behaviour of the game. In such circumstances, the errors resulting from an incorrect choice of expectation function are likely to be even smaller than those which appear in Table 5.1.

# The limitations of grading

Grades help tournament organizers to group players of approximately equal strength, and they provide the appropriate authorities with a convenient basis for the awarding of honorific titles such as 'master' and 'grandmaster'. However, it is very easy to become drunk with figures, and it is appropriate that this discussion should end with some cautionary remarks.

(a) Grades calculated from only a few results are unlikely to be reliable.

(b) The assumption underlying all grading is that a player's performance against one opponent casts light on his expectation against another. If this assumption is unjustified, no amount of mathematical sophistication will provide a remedy. In particular, a grade calculated only from results against much weaker opponents is unlikely to place a player accurately among his peers.

(c) There are circumstances in which grades are virtually meaningless. For an artificial but instructive example, suppose that we have a set of players in London and another set in Moscow. If we try to calculate grades embracing both sets, the placing of players within each set may be well determined, but the placing of the sets as a whole will depend on the results of the few games between players in different cities. Furthermore, these games are likely to have been between the leading players in each city, and little can be inferred from them about the relative abilities of more modest performers. Grading administrators are well aware of these problems and refrain from publishing composite lists in such circumstances, but players sometimes try to make inferences by combining lists which administrators have been careful to keep separate.

(d) A grade is merely a general measure of a player's performance relative to that of certain other players over a particular period. *It is not an absolute measure of anything at all.* The average ability of a pool of players is always changing, through study, practice, and ageing, but grading provides no mechanism by which the average grade can be made to reflect these changes; indeed, if the pool of players remains constant and every game causes equal and opposite changes to the grades of the affected players, the average grade never changes at all. What does change the average grade of a pool is the arrival and departure of players, and if a player has a different grade when he leaves than he received when he arrived then his sojourn will have disturbed the average grade of the other players; but this change is merely an artificial consequence of the grading calculations, and it does not represent any genuine change in average ability. It is of course open to a grading administrator to adjust the average grade of his pool to conform to any overall change in ability which he believes to have occurred, but the absence of an external standard of comparison means that any such adjustment is conjectural.

It is this last limitation that is most frequently overlooked. Students of all games like to imagine how players of different periods would have compared with each other, and long-term grading has been hailed as providing an answer. This is wishful thinking. Grades may appear to be pure numbers, but they are actually measures relative to ill-defined and changing reference levels, and they cannot answer questions about the relative abilities of players when the reference levels are not the same. The absolute level represented by a particular grade may not change greatly over a short period, but it is doubtful whether a player's grade ten years before his peak can properly be compared with that ten years after, and quite certain that his peak cannot be compared with somebody else's peak in a different era altogether. Morphy in 1857–8 and Fischer in 1970–2 were outstanding among their chess contemporaries, and it is natural to speculate how they would have fared against each other; but such speculations are not answered by calculating grades through chains of intermediaries spanning over a hundred years.[4]

# Cyclic expectations

Although cyclic results of the form '*A* beats *B*, *B* beats *C*, and *C* beats *A*' are not uncommon in practice, they are usually attributable to chance fluctuations. Occasionally, however, such results may actually

[4] Chess enthusiasts may be surprised that the name of Elo has not figured more prominently in this discussion, since the Elo rating system has been in use internationally since 1970. However, Elo's work as described in his book *The rating of chessplayers, past and present* (Batsford, 1978) is open to serious criticism. His statistical testing is unsatisfactory to the point of being meaningless; he calculates standard deviations without allowing for draws, he does not always appear to allow for the extent to which his test results have contributed to the ratings which they purport to be testing, and he fails to make the important distinction between proving a proposition true and merely failing to prove it false. In particular, an analysis of 4795 games from Milwaukee Open tournaments, which he represents as demonstrating the normal distribution function to be the appropriate expectation function for chess, is actually no more than an incorrect analysis of the variation within his data. He also appears not to realize that changes in the overall strength of a pool cannot be detected, and that his 'deflation control', which claims to stabilize the implied reference level, is a delusion. Administrators of other sports (for example tennis) currently publish only rankings. The limitations of these are obvious, but at least they do not encourage illusory comparisons between today's champions and those of the past.

represent the normal expectation. As light relief after the rigours of the last few sections, let us look at a few examples.

The rivers at Oxford and Cambridge are too narrow to permit boats to race abreast, so competitive rowing takes the form of 'bump' races. The boats start in a long line, and a boat which catches the one in front records a bump and takes its place next day. Now suppose that the leading boat $A$ consists of stayers, and the next boat $B$ of sprinters. It is now quite feasible that $B$ will catch $A$ early on the first day, but that $A$ will wear $B$ down and return the compliment late on the second. Provided that the third boat does not intervene, $A$ and $B$ now change places on alternate days. Such results do indeed occasionally happen.

This is not a true cycle, of course; the first boat has an inherent disadvantage in such a situation, and neither boat has shown itself strong enough to overcome this disadvantage. But now let us consider individual pursuit races around a small circular track, the runners starting at opposite sides and attempting to overtake each other (Figure 5.4). If we have three runners, $A$ being a long distance runner, $B$ a middle distance runner, and $C$ a sprinter, and they race against each other in pairs, then it is quite feasible that $A$ will escape $B$'s early rush and win in the long run, that $B$ will do the same to $C$, but that $C$ will be fast enough to catch $A$ early on.

**Figure 5.4** Catch as catch can

If this seems a little artificial, let us consider cross-country races between teams. The customary scoring system for such races gives one point to the first runner, two to the next, and so on, victory going to the team scoring the fewest points. Now suppose that we have nine runners $A, B, \ldots, I$ who are expected to finish in this order, and that $A$, $F$, and $H$ form team $X$, that $B$, $D$, and $I$ form team $Y$, and that $C$, $E$, and $G$ form team $Z$. If each runner performs exactly to expectation, the finishing orders in races between two teams are as

**Table 5.3** Cross-country running: the results of races between certain teams

| Runner | X against Y Place | Points X | Points Y | Y against Z Place | Points Y | Points Z | Z against X Place | Points Z | Points X |
|---|---|---|---|---|---|---|---|---|---|
| A | 1 | 1 |   | – |   |   | 1 |   | 1 |
| B | 2 |   | 2 | 1 | 1 |   | – |   |   |
| C | – |   |   | 2 |   | 2 | 2 | 2 |   |
| D | 3 |   | 3 | 3 | 3 |   | – |   |   |
| E | – |   |   | 4 |   | 4 | 3 | 3 |   |
| F | 4 | 4 |   | – |   |   | 4 |   | 4 |
| G | – |   |   | 5 |   | 5 | 5 | 5 |   |
| H | 5 | 5 |   | – |   |   | 6 |   | 6 |
| I | 6 |   | 6 | 6 | 6 |   | – |   |   |
| Total |   | 10 | 11 |   | 10 | 11 |   | 10 | 11 |

Team $X$ comprises runners $A$, $F$, and $H$; team $Y$, runners $B$, $D$, and $I$; team $Z$, runners $C$, $E$, and $G$. The runners are assumed always to finish in order $A \ldots I$.

shown in Table 5.3, and we see that $X$ wins against $Y$, $Y$ wins against $Z$, and $Z$ wins against $X$.

Such precise performances are unlikely in practice, but it is interesting that there are circumstances in which the apparently paradoxical results '$A$ beats $B$, $B$ beats $C$, and $C$ beats $A$' actually represent the normal expectation.

# 6

## BLUFF AND DOUBLE BLUFF

In this chapter, we consider games in which each player has a choice, and his object is to maximize a payoff which depends both on his own choice and on his opponent's. This is the simplest aspect of the 'theory of games' originally developed by John von Neumann.

In some games, such as chess, a player always knows everything that his opponent knows. In such a game, a player always has a single optimal choice (or, perhaps, a set of equivalent choices). Nobody may yet have discovered what this optimal choice actually is, but it can be proved to exist. We confine our attention to games in which one player knows something that the other does not: for example, the identity of a card which he has drawn from a pack. In such a game, there may not be a single optimal choice, but there is always an optimal strategy in some sense.

### I've got a picture

We start with a very simple card game. The first player draws a card at random from a pack, and bets either one or five units that it is a picture card (king, queen, or jack). The second player then either concedes this bet or doubles it. If he concedes, he pays the first player the amount of the bet, and the card is returned to the pack without being examined. If he doubles, the amount of the bet is doubled, the card is examined, and the doubled bet is paid accordingly.

At first sight, this is a bad game for the first player. The probability that he draws a picture card is only 3/13. Suppose that indeed he does so (Figure 6.1). He presumably bets five, his opponent presumably concedes, and he wins five. Alternatively, suppose that he fails to draw a picture. He now presumably bets one, his opponent presumably doubles, and he loses two. So he has a probability 3/13 of winning five and a probability 10/13 of losing two, leading to an average loss of 5/13.

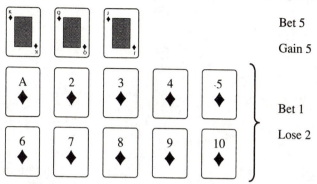

Bet 5

Gain 5

Bet 1

Lose 2

**Figure 6.1** 'I've got a picture': the obvious strategy for the first player

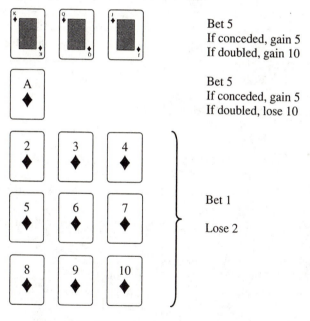

Bet 5
If conceded, gain 5
If doubled, gain 10

Bet 5
If conceded, gain 5
If doubled, lose 10

Bet 1

Lose 2

**Figure 6.2** 'I've got a picture': a better strategy

But now suppose that the first player bets five whenever he draws either a picture card *or an ace* (Figure 6.2). His probability of drawing one of these cards is 4/13, and if his opponent continues to concede all bets of five then he notches up a gain of five every time. Against this must be set the loss of two whenever he draws a card other than a picture or an ace, which happens with probability 9/13; but he still makes an average gain of 2/13.

After suffering this for a while, the second player is likely to realize that he is being taken for a fool and that at least some bets of five should be doubled. In practice, he is likely to study the first player's face in search of a clue, but let us assume that the first player is a skilled bluffer and that no clue is visible. In this case, the second player must rely on the laws of chance, and these state only that the probability that a bet of five represents a bluff on an ace is 1/4. Doubling gains ten in this case, but it loses ten whenever the card turns out to be a picture after all, and the latter happens with probability 3/4. So the average loss is still five.

So we see that the strategy of betting five on a picture card or an ace gives the first player an average expectation of 2/13 whatever his opponent may do. If all bets of five are conceded, fine. If some or all are doubled, the losses on the aces are balanced by the extra gains on the picture cards, and the overall expectation remains the same.

# An optimal strategy for each player

The strategy of betting five on a picture card or an ace is equivalent to the following:

(1a) on drawing a picture, always bet five;

(1b) on drawing any other card, bet five at random with probability 1/10, otherwise bet one.

The selection of the aces is no more than a convenient way of making a random choice with probability 1/10.

It can be shown that this represents the best possible strategy for the first player, in the sense that it generates the largest average expectation which can be guaranteed irrespective of the strategy subsequently adopted by the opponent. If he bluffs with probability less than 1/10, his opponent can reduce this average expectation by conceding all bets of five; if he bluffs with probability greater than 1/10, his opponent can reduce it by doubling all bets of five; and if he bluffs other than at random, his opponent may be able to work out what is coming and act accordingly. It is certainly true that other strategies may generate a larger expectation if the opponent plays badly; for example, betting five all the time shows a massive profit if the opponent never doubles at all. But it is prudent to allow for the best that the opponent can do, and bluffing at random with probability 1/10 leaves the opponent without effective resource.

Now let us look at matters from the other side. Consider the following strategy for the second player:

(2a) on hearing a bet of one, always double;

(2b) on hearing a bet of five, double at random with probability 7/15, otherwise concede.

Calculation shows that this strategy gives an average gain of two every time the first player draws a losing card, irrespective of whether the actual bet is one or five, but yields an average loss of 22/3 whenever the first player draws a picture. The probabilities of these cards are 10/13 and 3/13 respectively, so the overall average loss is 2/13. We have already seen that the first player can guarantee an average gain of 2/13 by following strategy (1a)/(1b), so the second player cannot hope to do better than this. Conversely, the fact that strategy (2a)/(2b) restricts the second player's average loss to 2/13 proves that the first player cannot achieve a greater average gain by varying from strategy (1a)/(1b).

But just as the second player can take advantage of any departure by the first from strategy (1a)/(1b), so the first player can take advantage of any departure by the second from strategy (2a)/(2b). If the second player doubles bets of five with probability less than 7/15, the first player can increase his expectation by bluffing on every card; if the second player doubles with probability greater than 7/15, the first player can increase his expectation by never bluffing at all.

Let us summarize this analysis. We have examined the best strategy for each player, we have seen how the opponent can take advantage of any departure from this strategy, and we have shown how the first player can ensure an average gain of 2/13 and how the second player can restrict his average loss to 2/13. This number 2/13 can therefore reasonably be called the 'value' of the game to the first player.[1]

---

[1] We have restricted the first player's choice to one or five units in order to simplify the discussion, but a more natural way to play this game is to allow him to bet any number of units from one to five inclusive. In the event, however, this extra freedom turns out to give no advantage. The second player's best strategy, on hearing a bet of $n$ units, is to double with probability $(n+2)/3n$. This gives the first player an average loss of two whenever he draws a losing card, while restricting his average gain to $(4n+2)/3$ whenever he draws a picture. So the first player should bet five on every picture card in order to maximize this gain, and his bluffs must conform.

# Scissors, paper, stone

The analysis of 'I've got a picture' is complicated by the fact that each player must prepare strategies for two eventualities: the first player because he may draw a winning or a losing card, and the second because he may hear a high or a low bet. Some aspects of the von Neumann theory are more easily demonstrated by a game in which a player need only consider one set of circumstances.

There is a well-known game for children in which each of two players holds a hand behind his back, chants a ritual phrase in unison, and brings his hand forward in one of three configurations: with two fingers in a V to represent scissors, with the whole hand spread out and slightly curved to represent paper, or with the fist clenched to represent a stone. The winner is determined by the rule that scissors cut paper, paper wraps stone, and stone blunts scissors. If both players display the same configuration, the game is drawn.

In this elementary form, the game is of no mathematical interest, but it can be given such interest by the introduction of differential scoring. Let us therefore suppose that scissors score 1 against paper, that paper scores 2 against stone, and that stone scores 3 against scissors. What is now the best strategy? Do we form a stone and hope to score 3, or do we expect our opponent to have formed a stone and try to beat him by forming paper, or what?

Plainly, if we always make the same choice then our opponent can make the appropriate reply and win, so we must mix two or more choices in a random manner. Let us therefore form scissors, paper, and stone at random with probabilities $x$, $y$, and $z$ respectively. This gives an average gain of $(3z - y)$ whenever our opponent forms scissors, $(x - 2z)$ whenever he forms paper, and $(2y - 3x)$ whenever he forms a stone. We cannot make all these gains positive, since this would require $3z > y$, $x > 2z$, and $2y > 3x$, from which it would follow that $6z > 2y > 3x > 6z$, which is impossible. So the best for which we can hope is that none of them is actually negative, and this can be achieved only by setting $x$ to 1/3, $y$ to 1/2, and $z$ to 1/6. It is perhaps unlikely that we would have guessed these probabilities without a detailed analysis.

Two points about this solution should be noted. If we vary from it in any way, our opponent can take advantage. For example, if we increase the frequency with which we form scissors relative to that with which we form paper, he can profit by always forming a stone.

Only by keeping strictly to the probabilities 1/3, 1/2, and 1/6 can we ensure that he can make no profit. On the other hand, if we do keep strictly to these probabilities, *it does not matter what our opponent does;* he can play his best strategy (which is the same as ours, since the game is the same for both players), or he can play any bad strategy that he likes (perhaps always forming scissors, or always paper, or always a stone), and our average expectation is zero in every case. In other words, we can ensure our optimal expectation only by forgoing any chance of profiting from our opponent's mistakes.

Similar effects occur in 'I've got a picture'. If the first player follows his best strategy, betting five on all winning cards and on one loser in ten chosen at random, he makes an average gain of 2/13 whether his opponent plays his own best strategy in reply to bets of five, or concedes them all, or doubles them all. He forgoes any chance of profiting should his opponent fail to double bets of five at the correct frequency. He *does* profit if his opponent fails to double a bet of one, however; such a choice is definitely inferior, and is to be avoided. In this respect, the game differs from 'scissors, paper, stone', which has no choice that should always be avoided. Similarly, if the second player doubles all bets of one and doubles bets of five at random with probability 7/15, he makes an average loss of 2/13 whether the first player bets five on losing cards at the correct frequency, or always, or never. He forgoes any chance of profiting should the first player fail to bet five on losing cards at the correct frequency, though he does profit if the first player bets less than five on a winning card.

# You cut, I'll choose

The equalization of an opponent's options is a strategy of wide applicability. The standard example is that of a parent who wishes to divide a cake between two children with the minimum of quarrelling, and tells one child to divide the cake into equal portions and then to give the other first choice. In practice, this 'game' is very slightly in favour of the second child, who can take advantage of any inequality in the cutting, but it is the first child's task to minimize this inequality and the second player's advantage is usually small.

A similar technique can be applied to any asymmetric task. Suppose that two labourers on a demolition site wish to settle an argument about their prowess in a gentlemanly way, but that only one edifice remains for destruction. It may not be practicable for them to demolish half each and see who finishes first, but it is always possible for one to challenge the other to demolish it within a certain time, and for the other either to attempt this task or to invite the challenger to do it himself. If the challenger believes the abilities of the candidates to be equal, his best strategy is to stipulate a target which he believes that he has only an even chance of achieving; any other stipulation allows his opponent to make an advantageous choice. If he believes the abilities to be unequal, he should stipulate a target at which he estimates that his own chance of success equals his opponent's chance of failure. *Note that this rule applies whether or not he regards himself as the stronger.* If he is indeed the stronger, it guarantees a certain chance of success (assuming that his estimate is indeed correct); if he is the weaker, it minimizes the amount that his opponent can gain.

In practice, of course, it is unusual for a task to be unrepeatable, and the usual procedure in an asymmetric game is for each player to take each role at least once. In a game of cricket, for example, each side bats in turn, the first innings being decided by the toss of a coin. But batting first may confer a distinct advantage (or disadvantage, depending on the circumstances), and there are occasions when winning the toss is almost tantamount to winning the game. This can be avoided by allowing the loser of the toss to state the side which he believes to have the advantage and to specify a bonus to be given to the other side, and for the winner then to decide whether to accept the bonus and concede the advantage or vice versa. This does not quite remove the advantage of winning the toss, because the loser may misjudge the bonus that should be given, but it greatly reduces it. A similar procedure can be applied to any game in which bonuses or handicaps can be assigned on a suitably fine scale. The bonus or handicap need not even be related to the score; for example, in a game such as chess, one player may be allowed longer for his moves than the other.

One final illustration may be of interest. In the climactic scene of an archetypal Western film, the hero and the villain appear at opposite ends of a deserted street, walk slowly towards each other, and fire simultaneously; and the villain drops dead. The death of the villain merely reflects the moral uprightness of motion picture makers, but the simultaneity is a matter of mathematics. Let the probability that hero and villain hit at distance $x$ be $h(x)$ and $v(x)$ respectively, and let us assume that a gun takes so long to reload that if either misses then his opponent can close in and finish him off with certainty. The best strategy for the hero is now to hold his fire until $h(x) = 1 - v(x)$.

If he delays longer, he increases the risk that he may be shot directly; if he fires earlier, he increases the risk that he may miss and be finished off at leisure. But by the same argument, the best strategy for the villain is to hold his fire until $v(x) = 1 - h(x)$, and this clearly yields the same value of $x$. I owe this delightful example to Peter Swinnerton-Dyer, who used to cite it in his lectures on the subject.

# The nature of bluffing

An inexperienced player thinks of a bluff as a manoeuvre whose objective is to persuade his opponent to play wrongly on a particular occasion. There are indeed many occasions on which such bluffs are successful, but their success depends upon the opponent's lack of sophistication. The bluffs which we have studied in this chapter are of a more subtle kind; their objective is not to persuade the opponent to take a wrong course of action on any particular occasion but to make his actual course of action irrelevant in the long term. These bluffs are effective however sophisticated he may be. They also avoid the need for 'bluff or double bluff' analyses such as 'I hold $A$ and want him to do $X$, so perhaps I should pretend to hold $B$, which will work if he swallows the bluff, or perhaps I should pretend to pretend to hold $A$, which will work if he thinks I'm bluffing, or perhaps . . . '. Look back at 'I've got a picture'. The best strategy for the first player, to bluff on every ace, makes it irrelevant in the long term whether the bluff is called or not.

Our bluffs in 'I've got a picture' are 'high' bluffs; we make the same high bet on a losing card as we make on winning cards. The inexperienced player is perhaps more likely to think in terms of 'low' bluffs (betting low on a winning card in order to lure the opponent into a trap), because such bluffs apparently cost less if they fail. Whether bluffing low is in fact the best strategy can be determined only by a detailed analysis of the pay-offs for the game in question, but very often it is not. In 'I've got a picture', for example, it is obvious that bluffing low on a winning card is disadvantageous; a low bet gains only two even if doubled, whereas a high bet gains five even if not doubled. But suppose that we change the rules and make the first player bet either four or five; does it not now pay the first player to bet low on a winning card, hoping to be doubled and score eight? No; the fact may be surprising, but it does not. It can be shown that the best strategy for the second player is now to double all bets of four, and to double bets of five with probability 13/15. This gains an average of eight whenever the first player holds a losing card, while conceding an average of either eight or 28/3 on a winning card according to whether the first player bets four or five. The latter is plainly the better for the first player, and in fact his best strategy is as before: to bet high on all winning cards, and on losing cards with probability 1/10.

On the other hand, there are occasions on which low bluffing is certainly correct. For example, if an opponent at bridge has embarked on a course of action which we can see to be misguided, we may pretend to weakness in order not to alert him while he still has time to change to something better. Many of the spectacular deceptions that are featured in books on bridge come into this category. There are even games in which the correct strategy is sometimes to lie completely doggo, making a fixed bet and giving no indication at all of the hand held; just as the best strategy for an animal or bird, caught in the open, may be to freeze and hope to be overlooked.

Even so, examples where low bluffs are profitable appear to be atypical, and to arise mainly in situations where we are bound to be in trouble if our opponent makes his best move. In most games that I have analysed, it is more profitable to bet high on all good hands and occasionally to bluff high than to bet low on all bad hands and occasionally to bluff low.

# Analysing a game

The actual obtaining of solutions (as opposed to demonstrating that a particular solution is in fact optimal) has been passed over rather lightly so far, but the time has come to say a little about it. A complete description is impracticable in a book of this kind, but the essence of the matter is quite simple: we expect a solution to contain a combination of our own options that equalizes the effects of our opponent's options, and the appropriate combination can be found by solving linear equations.

As an illustration, let us consider the game of Even Steven and the Odd Bod. This is played in a similar manner to 'scissors, paper, stone', but each player merely holds out one or two fingers, Steven winning if the total number of fingers on display is even and the Odd Bod if it is odd. The total itself determines the amount won, so if both players display two fingers then Steven wins four.

We start by choosing a player, and drawing up a table showing each of his options, each of his opponent's, and the amount that he wins in each case. Table 6.1 shows the amounts won by Steven. A complementary table can be constructed for the Odd Bod, but it is unnecessary because the whole analysis can be performed using Steven's.

**Table 6.1** The game of Even Steven and the Odd Bod: Steven's score

| The Odd Bod's choice | Steven's choice | |
|---|---|---|
| | 1 | 2 |
| 1 | 2 | —3 |
| 2 | —3 | 4 |

We now make the provisional assumption that Steven's optimal strategy involves choosing one finger with probability $x$ and two fingers with probability $y$, and that this strategy yields an average gain of $p$ whatever the Odd Bod does. If the Odd Bod displays one finger, the first row of Table 6.1 states that Steven's average gain is $2x-3y$, so we have

$$2x-3y=p. \ldots (1)$$

If the Odd Bod displays two fingers, the second row of the table states that Steven's average gain is $-3x+4y$, so we have

$$-3x+4y=p. \ldots (2)$$

Finally, $x$ and $y$ between them must cover all possibilities, so we have

$$x+y=1. \ldots (3)$$

This gives us three linear equations for three unknowns, and if we solve them in the usual way we find that $x=7/12$, $y=5/12$, and $p=-1/12$. Neither $x$ nor $y$ is negative, so we can realize this solution in actual play. It is not a good game for Steven.

It is now prudent to perform the same exercise for the Odd Bod, if only to confirm that his best strategy produces an average gain equal to Steven's loss. We wrote down Steven's equations by copying the rows of Table 6.1, and we can write down the Odd Bod's by copying the columns and negating the signs. This gives

$$-2x+3y=p \ldots (4)$$

and

$$3x-4y=p, \ldots (5)$$

and again we must have

$$x+ y=1. \ldots (6)$$

These equations have the solution $x=7/12$, $y=5/12$, and $p=1/12$. Again, neither $x$ nor $y$ is negative, so we can realize this solution in actual play, and the Odd Bod's average gain does indeed equal Steven's average loss.

So we have a complete analysis. Each player should choose one finger with probability 7/12 and two fingers with probability 5/12, and the average gain to the Odd Bod is 1/12.

If all games were as easy as this, there would be little to write about. In practice, however, there may be several problems.

(a) When the table of winnings is constructed, it may be found that some of the amounts are not simple numbers, but depend on unknown factors which must be averaged according to the laws of probability. This does not make any difference of principle, but it complicates the arithmetic.

(b) The number of options may be impracticably large. In poker, for example, a player has the option of conceding, or matching the current bet, or raising but conceding if raised again, or raising and matching if raised again, or raising twice and then conceding, or raising twice and then matching, and so on, and each of these options must be considered for every hand that he may hold.[2]

---

[2] I examined several simplified versions of poker while writing this chapter, but found none suitable for detailed exposition. Consider the following rules: (i) the pack is reduced to three cards, ranked king (high), queen, jack; (ii) each player bets one unit, draws one card, and examines it; (iii) the first player either bets a further unit or concedes; (iv) if the first player bets, the second player must either concede, or match the bet, or bet one more; (v) if the second player bets one more, the first player must either match the new bet or concede; (vi) if nobody has conceded, the higher card wins. A game could hardly be simpler and still preserve the essentials of poker, but a full analysis of even this simple game is too lengthy to give here. For the curious, the optimal strategies are as follows. For the first player: holding the king, always bet, and always match a raise; holding the queen, always bet, but match a raise only with probability 1/5; holding the jack, bet only with probability 1/3, and always concede a raise. For the second: holding the king, always raise; holding the queen, match with

(c) Although the equations may be soluble in theory, their solution may involve computational difficulties. Exact computation may generate fractions with unacceptably large denominators, while the replacement of exact fractions such as 1/3 by approximations such as 0.3333 may significantly change the solution.

(d) Even in theory, the equations may not have a solution at all, or their solution may require a particular choice to be made with negative probability.

From the point of view of abstract mathematics, as opposed to practical computation, the last of these problems is the most important, and it is to this that the theory of von Neumann addresses itself. The crucial theorem is as follows: it is always possible for each player to select a *subset* of the options available to him, and to make a combination of choices within this subset which gives (i) the same result if the opponent chooses an option within his own subset, and (ii) at least as good a result if his opponent chooses any other option. For simple games such as 'I've got a picture', these subsets can be determined in an *ad hoc* manner. More complicated games demand systematic refinement of trial subsets; we select a subset for each player, solve the resulting equations, look at the remaining options, see if either player can obtain a better result by choosing such an option, modify his subset if he can, and continue until a solution is found. Detailed algorithms can be found in textbooks on computation. For games with moderate numbers of options, the task is in fact within the capabilities of a modern home computer, but the writing of a suitable program is not a job for an amateur.

In any case, little is really gained by solving games of this kind on a computer. Yes, it tells us the numerical answers, but it does so only for rather artificial games. Real-life versions of poker are far too complicated to be analysed on contemporary computers. Perhaps you have visions of discovering a game with an obscure strategy and an unexpected average gain, on which you can found a career as a gambler. If you do have such ambitions, you will find that your primary need is not for a computer or a book on mathematical games; your need is for Big Louie to safeguard the take, Slinky Lulu to attract the punters, and other estimable henchpersons.

Which brings us to the final point. Do not think that a reading of

---

probability 1/3, otherwise concede; holding the jack, raise with probability 1/5, otherwise concede. The second player's strategy guarantees an average gain of 13/90, and the first player's holds him to this.

this chapter has equipped you to take the pants off your local poker school. Three assumptions have been made: that you can bluff without giving any indication, that nobody is cheating, and that the winner actually gets paid. You will not necessarily be well advised to make these assumptions in practice.

# 7

# THE ANALYSIS OF PUZZLES

We now proceed to games of pure skill, and we start by looking at some of the mathematical techniques which can be used when solving puzzles. There is no general way of solving puzzles, of course; if there were, they would cease to be puzzling. But there are a number of techniques which frequently help either to solve a puzzle or to prove that no solution exists.

## Black and white squares

By far the most useful of these techniques is the exploitation of parity. It usually occurs when a puzzle involves moving things around, and the possible configurations can somehow be divided into 'odd' and 'even' classes. If there is no way of moving from one class into the other, the analysis of parity is purely a negative technique, and is used to prove insolubility; but it may also be useful when a small number of ways of moving from one class to the other do exist, since it may point the need for a solution to incorporate one of them.

Perhaps the most familiar examples of parity occur on the chessboard. Every beginner knows that the bishop, which can only move diagonally, is permanently restricted to squares of the colour on which it starts (Figure 7.1, left). More subtle is the case of the knight, which moves from a white square to a black (Figure 7.1, right) or vice versa. If two squares are of the same colour, a knight can move from one to the other in an even number of moves, but never in an odd number.

Most chess problems involve specialized and recondite tactics which are outside the scope of this book, but Figure 7.2 shows a simple puzzle in which parity plays a decisive role. The position shown in this figure differs from the initial array only in that one White pawn has advanced by one square, and the problem is to discover in how few moves this position could have been reached in play. If it is now Black's move, there is no difficulty; White could simply have moved

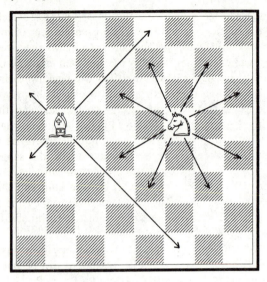

**Figure 7.1** Bishop and knight

**Figure 7.2** Chess: how quickly can we reach this position?

his pawn. But suppose we are told that it is *White's* move? However he has played, Black must have made an even number of moves; his knights are now on squares of opposite colours, as they were at the start of the game, so they have made an even number of moves between them (irrespective of whether they are back on their original squares or have changed places), and his rooks, if they have moved at all, can only have moved one square to the side and then back again. By the same argument, White's knights and rooks have made an even number of moves between them; but his pawn has made an odd number of moves, and he must have made the same overall number of moves as Black, so some other man must have made an odd number also. The only candidates are the king and the queen, and the shortest possibility consists of seven moves by the king. So to reach Figure 7.2 with White to play must have involved at least eight moves by each side.

But if the typical chess problem is too specialized for inclusion here, the black-and-white colouring of the chessboard can be put to good use in other contexts. We give three examples.

(a) The domino problem. This is a classic swindle. The victim is given a board shaped as in Figure 7.3, and is told to cover it with 31 dominoes. He is likely to struggle until it occurs to him to colour alternate squares black and white, upon which he discovers that he has 32 black squares but only 30 white ones. Since a domino must always cover one square of each colour, the problem is insoluble.

(b) Polyominoes. These are generalizations of dominoes, and comprise squares joined together in various ways. There are two

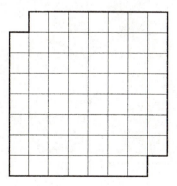

**Figure 7.3** The domino problem

different 'trominoes' with three squares each, and five 'tetrominoes' with four squares (Figure 7.4). The five tetrominoes between them occupy 20 squares, and it is natural to ask if they can be arranged so as to fill a $4 \times 5$ rectangle. It turns out that they cannot, and the method of parity provides an elegant proof. Let alternate squares of the rectangle be coloured black and white. We now see that the 'T' tetromino always covers three black squares and one white (or vice versa), whereas the other four tetrominoes always cover two squares of each colour. So the five tetrominoes between them must cover an odd number of squares of each colour, whereas we need to cover ten of each. On the other hand, if we add the two trominoes and a domino, we can cover a $4 \times 7$ rectangle; the trominoes each cover two squares of one colour and one of the other, and can be placed so as to balance the 'T' tetromino.[1]

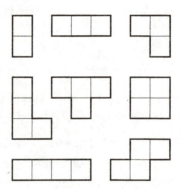

**Figure 7.4** One domino, two trominoes, five tetrominoes

(c) The 'fifteen' puzzle. This puzzle, due to Sam Loyd, is another classic. The victim is given a tray which contains fifteen numbered blocks and one blank space, and is told to reverse the order of the blocks by shuffling (Figure 7.5). The crucial property of this problem results from the fact, proved in textbooks on group theory, that a permutation is either 'odd' or 'even' depending on the number of interchanges needed to achieve it; an odd permutation can never be achieved by an even number of interchanges, nor vice versa. The

[1] Polyominoes are discussed at length in *Polyominoes* by Solomon W. Golomb (Allen and Unwin, 1966). There are 12 pentominoes, which can be arranged in a rectangle; and 35 hexominoes, which cannot.

| Start | | | |
|---|---|---|---|
| 1 | 2 | 3 | 4 |
| 5 | 6 | 7 | 8 |
| 9 | 10 | 11 | 12 |
| 13 | 14 | 15 | |

| Target 1 | | | |
|---|---|---|---|
| | 15 | 14 | 13 |
| 12 | 11 | 10 | 9 |
| 8 | 7 | 6 | 5 |
| 4 | 3 | 2 | 1 |

| Target 2 | | | |
|---|---|---|---|
| 15 | 14 | 13 | 12 |
| 11 | 10 | 9 | 8 |
| 7 | 6 | 5 | 4 |
| 3 | 2 | 1 | |

**Figure 7.5** The 'fifteen' puzzle

middle position in Figure 7.5 can be obtained from the starting position by interchanging 1 and the blank, 2 and 15, 3 and 14, 4 and 13, 5 and 12, 6 and 11, 7 and 10, and 8 and 9, so it is an even permutation of the starting position. On the other hand, the right-hand position can be obtained from the starting position by interchanging 1 and 15, 2 and 14, 3 and 13, 4 and 12, 5 and 11, 6 and 10, and 7 and 9, so it is an odd permutation. But the basic move in the puzzle interchanges the blank space with one of the blocks next to it, and if we colour the spaces as on a chessboard, we see that each such interchange moves the blank to a space of the opposite colour; to restore it to a space of the original colour takes an even number of moves. The blank spaces in the target positions of Figure 7.5 are on squares of the same colour as that in the starting position, so these positions can be reached only in even numbers of moves. It follows that the right-hand position cannot be reached, though the middle one can.[2]

The placing of trominoes so as to balance the 'T' tetromino within a 4 × 7 rectangle provides an example of the use of parity to point the way to a solution, but the problem in question is so simple that a solver may be expected to succeed without this aid. A better example is provided by the 'fifteen' puzzle. Suppose that the numbers are replaced by letters, two of which are the same. Any required pattern can now be achieved, since an apparently odd permutation can be made into an even one by interchanging the identical letters. My favourite version is DEAD PIGS WON'T FLY (Figure 7.6, left). The

---

[2] This is one of the puzzles discussed in *Sliding piece puzzles* by Edward Hordern, which is another volume in the present series. Other puzzles discussed in the series are Rubik's cube (in *Rubik's cubic compendium* by Ernö Rubik and others) and solitaire (in my own *The ins and outs of peg solitaire*). These are all classic puzzles and it would be wrong for this book to ignore them entirely, but our brief treatment reflects the fact that fuller information is readily available elsewhere.

**Figure 7.6** A problem in porcine levitation

procedure is to show this legend to the victim and then to invite him to produce DEAD PIGS FLY TOWN (Figure 7.6, right). If he does not know the trick, he will not think of disturbing what is apparently already correct and will get no further than DEAD PIGS FLY TONW. The effect is enhanced if blocks of a different colour are used for each word; a purist may deplore such colouring as an irrelevant garnish, but it makes the victim even more reluctant to disturb the part of the layout that is apparently correct.

# Divisibility by three

Having seen the power of arguments involving parity, it is natural to ask if divisibility by other numbers provides a similarly powerful weapon.

Occasionally, it does, though such effects are not as common as those involving parity. An example is given by the problem of placing $n$ non-interfering queens on a cylindrical chessboard. A queen can move along any straight line, horizontally, vertically, or diagonally, and the classic 'eight queens' problem requires the placing of eight such queens on an ordinary chessboard such that none impedes the move of any other. This is easily generalized to the case of $n$ queens on an $n \times n$ board, and we shall consider it later in the chapter. For the present, however, we consider an elegant variation in which the board is bent into a cylinder (Figure 7.7). Such a board has only $n$ diagonals in each direction, since a diagonal which apparently reaches the side of the board simply reappears on the other side and continues. Figure 7.7 shows the two sets of diagonals, $ABC\ldots$ in one direction and $abc\ldots$ in the other. The problem therefore reduces to the following: Place $n$ queens on an $n \times n$ cylindrical board so that there is one and only one in each row, column, $ABC$ diagonal, and $abc$ diagonal.

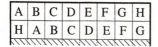

**Figure 7.7** Diagonals on a cylinder

Experiment soon shows the problem to be insoluble on a $2 \times 2$, $3 \times 3$, or $4 \times 4$ board, but to be soluble on a $5 \times 5$ board by a line of knight's moves (Figure 7.8, left). The $6 \times 6$ board again proves to be insoluble, but the $7 \times 7$ yields to another line of knight's moves (Figure 7.8, right). A pattern now begins to emerge. A line of knight's moves automatically ensures that we have one queen in each row and in each *ABC* diagonal. If *n* is odd, it also ensures that we get one queen in each column; but if *n* is even, half the columns receive two queens and the rest receive none (Figure 7.9, left). Similarly, if *n* is not divisible by 3, we get one queen in each of the *abc* diagonals, but if *n* is divisible by 3, a third of them receive three queens and the rest receive none (Figure 7.9, right). So a line of knight's moves solves the problem if and only if *n* is divisible by neither 2 nor 3. The same can be shown to be true of other arrangements in straight lines; they work if *n* is divisible by neither 2 nor 3, but fail otherwise.

But we are not restricted to arrangements in straight lines. Are there no arrangements that work when straight lines fail?

It turns out that there are not; and the proof, although a little harder than most of the material in this book, is sufficiently elegant to be worth giving in full. Let us start with the case of even *n*, which is the easier. We number everything from 0 to $n-1$ in an obvious

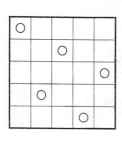
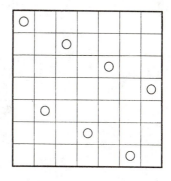

**Figure 7.8** Successful arrangements of queens on a cylinder

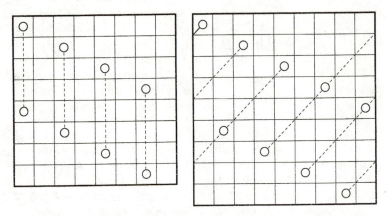

**Figure 7.9** Unsuccessful arangements of queens on a cylinder

way, rows from top to bottom, columns from left to right, and diagonals from $A$ or $a$ as appropriate; and we use the result, proved in textbooks on algebra, that $0+1+ \ldots +(n-1)=n(n-1)/2$. Now suppose that the queen in row $i$ is in column $q_i$. The numbers $q_0, q_1, \ldots, q_{n-1}$ must be a permutation of the numbers $0, 1, \ldots, n-1$, since no two queens lie in the same column, so

$$q_0 + q_1 + \ldots + q_{n-1} = 0 + 1 + \ldots + (n-1)$$
$$= n(n-1)/2.$$

Now let us consider the $ABC$ diagonals. Suppose that the queen in row $i$ is in diagonal $r_i$; then either

$$r_i = q_i - i$$

or

$$r_i = q_i - i + n$$

according as $q_i \geqslant i$ or $q_i < i$. Suppose that the longer equation $r_i = q_i - i + n$ applies to $k$ of the diagonals; then

$$r_0 + r_1 + \ldots + r_{n-1} = (q_0 + q_1 + \ldots + q_{n-1}) - \{0 + 1 + \ldots + (n-1)\} + kn.$$

But $q_0 + q_1 + \ldots + q_{n-1}$ and $0 + 1 + \ldots + (n-1)$ both equal $n(n-1)/2$, so we are left with

$$r_0 + r_1 + \ldots + r_{n-1} = kn.$$

However, the numbers $r_0, r_1, \ldots, r_{n-1}$ must be another permutation of

$0,1, \ldots, n-1$, so the sum $r_0 + r_1 + \ldots + r_{n-1}$ must be $n(n-1)/2$, whence $k$ must equal $(n-1)/2$; and if $n$ is even, $(n-1)/2$ is not an integer, which is a contradiction. So the problem is insoluble if $n$ is even.

The case where $n$ is divisible by 3 can be analysed similarly. We now consider the squares of the relevant numbers, making use of the fact that $0^2 + 1^2 + \ldots + (n-1)^2 = n(n-1)(2n-1)/6$; and we need to take account of the *abc* diagonals, which involves the use of numbers $s_i$ defined by

$$s_i = q_i + i$$

or

$$s_i = q_i + i - n$$

as appropriate. A little manipulation now shows that

$$r_i^2 + s_i^2 = 2q_i^2 + 2i^2 - k_i n,$$

where $k_i$ is an integer which depends on the precise equations satisfied by $r_i$ and $s_i$; and if we add up these equations for $i = 0, 1, \ldots, n-1$, we get

$$(r_0^2 + r_1^2 + \ldots + r_{n-1}^2) + (s_0^2 + s_1^2 + \ldots + s_{n-1}^2)$$
$$= 2(q_0^2 + q_1^2 + \ldots + q_{n-1}^2) + 2\{0^2 + 1^2 + \ldots + (n-1)^2\} - kn$$

for some new integer $k$. However, the numbers $q_0, q_1, \ldots, q_{n-1}$, $r_0, r_1, \ldots, r_{n-1}$, and $s_0, s_1, \ldots, s_{n-1}$ are all permutations of $0, 1, \ldots, n-1$, so all the sums of squares are equal to $n(n-1)(2n-1)/6$, and it follows that $k$ must equal $(n-1)(2n-1)/3$; but $(n-1)(2n-1)/3$ is not an integer if $n$ is a multiple of 3, so again we have a contradiction. Hence the problem is insoluble under this condition also.

So it is possible to arrange $n$ non-interfering queens on an $n \times n$ cylindrical board only if $n$ is divisible by neither 2 nor 3; which is a simple and elegant result. I am told that it dates back to Euler.

# Positions with limited potential

Many problems in applied mathematics can be resolved by considering potentials. For example, when a pendulum is at the top of its swing, it has a certain potential energy, and it will attain a certain speed by the time that it reaches the bottom. Unless it receives extra energy from somewhere, it cannot attain more than this speed, nor can it swing through to a point higher than its starting point.

Similar arguments can sometimes be put to good use in the solution of puzzles. For example, the permitted move in a puzzle such as peg solitaire is to jump one peg over another and remove the peg jumped over, so if $P$, $Q$, $R$ are three holes in line, a move from $P$ over $Q$ into $R$ removes pegs from $P$ and $Q$ and puts a peg into $R$. If we now assign values $p$, $q$, $r$ to the holes and add up the values of the holes which are occupied, the effect of the move is to replace $p+q$ by $r$; and if $p+q \geqslant r$, this change is not an increase. So if we can assign values to the holes such that $p+q \geqslant r$ for *every* set of three adjacent holes in line, we have a measure of the 'potential' of a position; no move can increase such a potential, so a problem is insoluble if the target position has a higher potential than the starting position.

This simple fact allows an elegant resolution of certain problems, a striking example being given by Conway's problem of the solitaire army. Suppose we have an indefinitely large board with a line across it (Figure 7.10). If we place pegs only below this line and try to jump a man forward, how far forward can we go? It is not difficult to see that we can go forward up to four rows; the various sections of Figure 7.10 show possible starting configurations, and the actual moves are easy. But to reach the next row proves not to be so straightforward.

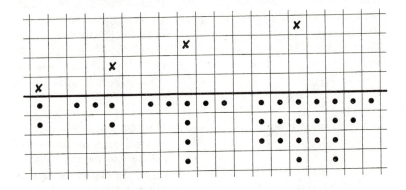

**Figure 7.10** Solitaire detachments

The reason is given by the values shown in Figure 7.11. Our target is the hole with value 1, and $\phi = (\sqrt{5}-1)/2$. This value $\phi$ has the property that $\phi^2 + \phi = 1$, so the relation $p+q \geqslant r$ is satisfied for every

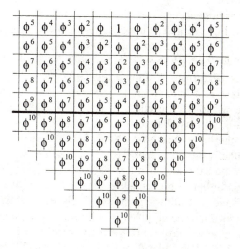

**Figure 7.11** The resolution of the solitaire army

set of adjacent holes $P$, $Q$, $R$ in line, and so no move can increase the sum of the occupied holes. Now suppose that all our pegs are initially below the thick line, and that the most distant hole actually occupied has value $\phi^n$. If we look at the first row below the thick line, and consider its centre hole and the holes to the left thereof, we see that the sum of the occupied holes cannot exceed $\phi^5 + \phi^6 + \phi^7 + \ldots + \phi^n$. This is a geometric series of positive terms with constant ratio less than 1, and a standard theorem of algebra states that the sum of such a series is always less than $a/(1-r)$, where $a$ is the first term and $r$ the constant ratio; so the sum of this particular series is less than $\phi^5/(1-\phi)$. But $1-\phi=\phi^2$, so this means that the sum is less than $\phi^3$. By a similar argument, the sum of the occupied holes to the right of the central hole is less than $\phi^4$; and $\phi^3 + \phi^4 = \phi^2$, so the total sum of the occupied holes in the row is less than $\phi^2$. Similarly, the sum of the occupied holes in the next row is less than $\phi^3$, and so on down to the lowest row that is occupied. A similar argument can now be applied to the numbers $\phi^2$, $\phi^3$, $\ldots$, that bound the row sums, thus proving that the sum of these bounds is less than 1. So however many holes we occupy, the sum of their values is less than 1; and it follows that we can never play a man into the target hole.

Solitaire offers many lovely problems, as those who read *The ins and outs of peg solitaire* will discover, but this is one of the loveliest.

# Systematic progress within a puzzle

Our final technique is that of systematic progress. It can take two forms: progress within a single puzzle, and progress from one puzzle to another.

An excellent example of progress within a single puzzle is provided by Rubik's cube. This puzzle is now well known, but a brief discussion is in order since similar techniques can be applied elsewhere.

The fundamental problem with Rubik's cube is that a single physical operation (rotating one face) does something logically complicated (moving several constituent cubelets). What we would prefer to have is a set of operations that may be physically complicated but are logically simple; for example, interchanging a single pair of cubelets while leaving everything else unchanged. In practice, we cannot quite achieve this; we cannot interchange one pair of cubelets in isolation, nor rotate a single corner or edge cubelet without rotating at least one other cubelet of the same type. But we can interchange two corner cubelets without disturbing other *corners*, since the necessary balancing interchange can be performed on edge cubelets. For proofs of all these statements, see *Rubik's cubic compendium*.

In fact, a suitable set of elementary operations is that shown in Figures 7.12 and 7.13. That this set does indeed suffice is shown by the following argument.

(a) Starting with a scrambled cube, and taking the face centres as reference points, we can use Figure 7.12(a) to get all the corners into position. We ignore their orientations at this stage.

(a)                                                (b)

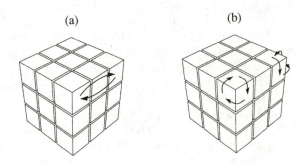

**Figure 7.12** Rubik's cube: elementary corner operations (possibly disturbing edges)

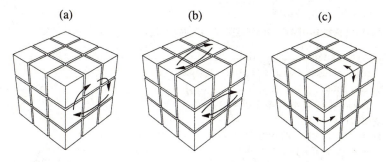

**Figure 7.13** Rubik's cube: elementary edge operations (preserving corners)

(b) Having positioned the corners, we can use Figure 7.12(b) to orient at least seven of them correctly. The correct orientation of the eighth corner should now be automatic; if it isn't, the cube has been badly assembled and must be taken apart and corrected. This is not usually a problem in practice, because manufacturers build uncoloured cubes and then colour the faces. Gentlemen do not take other people's cubes apart and reassemble them wrongly; it isn't really funny, and the risk of damage is appreciable.

(c) We can now choose any face, and use Figures 7.13(a) and 7.13(b) to put its edges in position. Figure 7.13(a) is sufficient to get all the edges on to the face, and to position at least one opposite pair correctly; and if the second pair is not positioned correctly in the process, Figure 7.13(b) can be used to interchange them.

(d) Similarly, we can position the edges correctly on a second face adjacent to the first, and then on a third face opposite to either of the first two. Figure 7.13(b) disturbs a face other than that being worked on, but it is always possible to ensure that this is not a face which has already been arranged. This deals with ten of the twelve edges, and if the remaining two do not automatically move to their correct positions in the process then the cube has again been wrongly assembled.

(e) Finally, we can use Figure 7.13(c) to orient at least eleven of the edges correctly; and if the twelfth does not become correctly oriented in the process, the cube has once more been wrongly assembled.

This procedure is by no means optimal in terms of the number of operations needed to restore a cube from a typical state of disarray, but at least it gets the thing out, and it provides a sound basis for further exploration. Solutions to Figures 7.12 and 7.13 are postponed until later in the chapter, in case you have previously been bewildered by the cube but now feel like trying again. Figure 7.13(b) is not strictly necessary as a member of the basic set, since the same effect can be achieved by six suitably chosen performances of Figure 7.13(a), but it is convenient to regard it as an independent operation.

## Systematic progress between puzzles

The value of systematic progress from one puzzle to another is not always so obvious. If a set of puzzles is presented as a series, it is indeed natural to look for ways of deriving each solution from the previous one; but a puzzle is often presented in isolation, and the opportunity for a systematic approach may be overlooked.

A simple example is provided by the problem of placing eight non-interfering queens on a chessboard. This problem can be solved in isolation, but the generalization to an $n \times n$ board is instructive.

We considered this problem earlier on a cylinder, and found that there are solutions if and only if $n$ is divisible by neither 2 nor 3. These solutions (Figure 7.8, for example) are clearly valid for a flat board as well. This deals with all values of $n$ of the form $6k \pm 1$. Furthermore, if a solution on a flat $n \times n$ board has a queen in the top left-hand corner, we can obtain a solution for an $(n-1) \times (n-1)$ board by omitting the first row and column. The solutions in Figure 7.8 do have such a queen, and the corresponding reduced solutions are shown in Figure 7.14. This deals with all values of $n$ of the form $6k - 2$ or $6k$.

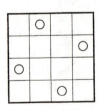
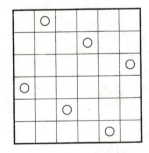

**Figure 7.14** Four and six queens

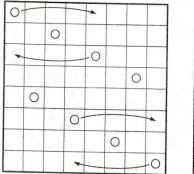 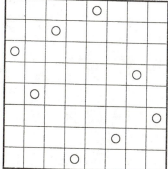

**Figure 7.15** Eight queens

There remain values of the form $6k+2$ and $6k+3$. Now if we take $n=8$ as an example and lay out parallel lines of knight's moves from opposite corners (Figure 7.15, left), we find that there is only one clash; the top left-hand and bottom right-hand queens are on the same diagonal. This can be cured by interchanging columns as shown, producing the position shown on the right-hand side of Figure 7.15. Furthermore, this technique can be generalized; if $n=6k+2$, we can always obtain a solution by laying out two parallel lines of knight's moves, taking the first and the penultimate queen in each line, and interchanging their columns. Figure 7.16 shows this process applied to a $14 \times 14$ board.

This leaves only values of the form $6k+3$, and for these we can apply the reverse of the procedure which we used for values of the forms $6k-2$ and $6k$. Our solutions for $n=6k+2$ never place a queen on the leading diagonal, so we can always add an extra row at the top and an extra column on the left, and place a queen in the new top left-hand corner.

So, by considering six separate classes, we have proved the problem soluble on all boards from $4 \times 4$ upwards.

The most formal of all systematic procedures is that of successive induction. This is often used as a method of proof; if a proposition is true for an initial integer $n_0$, and if its truth for $n$ implies its truth for $n+1$, then it must be true for all integers from $n_0$ upwards. This is a very powerful method of proof, and it can be just as powerful as a method of solution.

For an example, consider the classic 'twelve coin' problem. We are given twelve coins, one of which is counterfeit and does not weigh

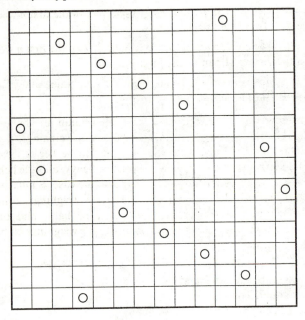

**Figure 7.16** Fourteen queens

the same as a true coin, and we have three weighings in which to identify the defaulter and to determine whether it is heavy or light. This is usually presented as an isolated puzzle, but it yields instructively to induction.

Let us suppose that we can identify a single counterfeit among $c$ coins in $n$ weighings in such a way that no coin occupies the same pan in every weighing. If we follow the procedure below, we can now identify a single counterfeit among $3c+3$ coins in $n+1$ weighings, again with no coin occupying the same pan in every weighing.

(a) We take coins 1 to $c$, distribute them among the first $n$ weighings as in the known solution, and omit them from the final weighing.

(b) We take coins $c+1$ to $2c$, distribute them similarly among the first $n$ weighings, and put them in the left-hand pan in the final weighing.

(c) We take coins $2c+1$ to $3c$, do the same for the first $n$ weighings, and put them in the right-hand pan in the final weighing.

(d) We place coin $3c+1$ in the left-hand pan in the first $n$ weighings, and omit it from the final weighing.

(e) We place coin $3c+2$ in the right-hand pan in the first $n$ weighings and in the left-hand pan in the final weighing.

(f) We omit coin $3c+3$ from the first $n$ weighings, and place it in the right-hand pan in the final weighing.

It is now easily seen that we have a solution to the extended problem. Suppose that the results of the first $n$ weighings, if applied to the first $c$ coins only, would show coin $j$ to be heavy. They now show that either coin $j$, $c+j$, or $2c+j$ must be heavy, and the behaviour of the final weighing tells us which in fact is the case. Similarly, if the left-hand pan tips down in each of the first $n$ weighings, either coin $3c+1$ must be heavy or coin $3c+2$ must be light, and the final weighing tells us which is which; and if the first $n$ weighings all balance,

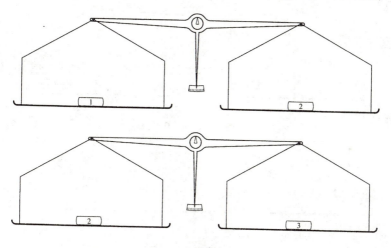

**Figure 7.17** Three coins, two weighings

**Table 7.1** Analysis of the weighings shown in Figure 7.17

| Low pan | | | Low pan | | | Low pan | | |
| 1 | 2 | Diagnosis | 1 | 2 | Diagnosis | 1 | 2 | Diagnosis |
|---|---|---|---|---|---|---|---|---|
| L | L | impossible | — | L | 3 light | R | L | 2 heavy |
| L | — | 1 heavy | — | — | impossible | R | — | 1 light |
| L | R | 2 light | — | R | 3 heavy | R | R | impossible |

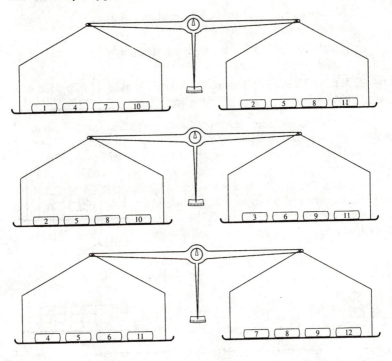

**Figure 7.18** Twelve coins, three weighings

**Table 7.2** Analysis of the weighings shown in Figure 7.18

| Low pan 1 | 2 | 3 | Diagnosis | Low pan 1 | 2 | 3 | Diagnosis | Low pan 1 | 2 | 3 | Diagnosis |
|---|---|---|---|---|---|---|---|---|---|---|---|
| L | L | L | impossible | — | L | L | 9 light | R | L | L | 5 heavy |
| L | L | — | 10 heavy | — | L | — | 3 light | R | L | — | 2 heavy |
| L | L | R | 11 light | — | L | R | 6 light | R | L | R | 8 heavy |
| L | — | L | 4 heavy | — | — | L | 12 light | R | — | L | 7 light |
| L | — | — | 1 heavy | — | — | — | impossible | R | — | — | 1 light |
| L | — | R | 7 heavy | — | — | R | 12 heavy | R | — | R | 4 light |
| L | R | L | 8 light | — | R | L | 6 heavy | R | R | L | 11 heavy |
| L | R | — | 2 light | — | R | — | 3 heavy | R | R | — | 10 light |
| L | R | R | 5 light | — | R | R | 9 heavy | R | R | R | impossible |

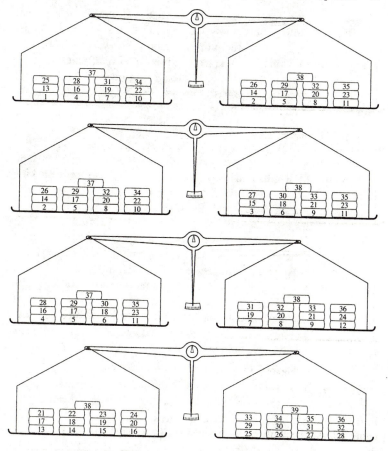

**Figure 7.19** Thirty-nine coins, four weighings

coin $3c+3$ must be the counterfeit, and the final weighing tells us whether it is light or heavy.

It remains only to find a solution for some initial value of $n$, and this is easily done. If we have three coins, the arrangement shown in Figure 7.17 identifies the defaulter in two weighings; Table 7.1 supplies the analysis. (The pattern of Figure 7.17 follows that of steps (d) to (f) of the systematic procedure, and can be regarded as the result of applying this procedure to the even simpler 'solution' in which the defaulter among no coins is identified in one weighing by placing no coins on each side: an amusing sophistry, though hardly of practical importance.) We can now apply the procedure to solve twelve coins in three weighings, giving the pattern which is shown in Figure 7.18 and analysed in Table 7.2. The next step is to solve 39 coins in four weighings, which can be done as shown in Figure 7.19, and we can continue this as long as we wish; given $n$ weighings, we can identify a single counterfeit from among $(3^n - 3)/2$ coins.[3]

To complete our study of puzzles, we give the promised solution to Rubik's cube. We assume the cube to be held so that one face is immediately opposite the eye, and we denote the faces by Front and Back, Left and Right, and Top and Bottom. We use the following operations: (i) rotations of the front face, denoted by $F_C$ (quarter turn clockwise), $F_H$ (half turn), and $F_A$ (quarter turn anticlockwise); (ii) rotations of the top face, denoted by $T_L$ (quarter turn left), $T_H$ (half turn) and $T_R$ (quarter turn right); (iii) quarter turns left of the bottom face and of the whole cube, denoted by $B_L$ and $W_L$ respectively; and (iv) quarter turns of the middle slice between the left and right faces, denoted by $M_U$ (up) and $M_D$ (down). In all these rotations, the direction 'left', 'right', 'up', or 'down' refers to the movement of the front edge. The slice and whole cube movements disturb the face centres, but they are always used in cancelling pairs or in complete sets of four, so the original centres are restored when the sequence finishes.

Figure 7.12(a). Hold the cube so that the corners to be interchanged are at the top of the front face, and perform the operations $F_A.W_L$ four times, then $T_R$ once (nine rotations in all, including four whole cube rotations). If the final operation $T_R$ is omitted, the corners of the top face are cycled (front right to back left to back right). This simple sequence gives a one-third twist to each of the corners that are not interchanged. If you dislike these twists, you can get rid of them by performing the sequence three times, but there is no real need to do so since we are not yet attempting to produce correct orientations.

Figure 7.12(b). Hold the cube similarly, and perform the operations $T_L.F_A$ three times, then $T_R.F_C$ three times (twelve rotations in all). The former top faces of the affected corners come on to the front face.

Figure 7.13(a). Hold the cube so that the edges to be cycled are on

---

[3] It is possible to renumber the coins so that the dispositions of coin $j$ can be written down immediately from the representation of $j$ in ternary arithmetic. Some readers may wish to work out the details; others will find them in an excellent paper by F. J. Dyson in *Math. Gazette* **30** (1946), 231–4.

the front face, left to top to right; then perform the operations $M_U.F_C.M_D.F_H.M_U.F_C.M_D$ (seven rotations in all, including four slice rotations). The up-down-up-down pattern of the slice moves contrasts with the unidirectional rotation of the front face. This sequence reorients two of the cycled edges; a more complicated version exists which avoids this, but again there seems little point in using it since we are not yet attempting to produce correct orientations. If the two operations $F_C$ are replaced by $F_A$, the direction of the cycle is reversed.

Figure 7.13(b). Hold the cube so that the edges to be interchanged are on the top and front faces, left to right in each case; then perform the operation $T_H.F_H$ three times (six rotations in all).

Figure 7.13(c). Hold the cube so that the edges to be reoriented are on the front face, top and left; then perform the operation $F_C$, then $M_D.B_L$ four times, then $F_A$, then $M_D.B_L$ four times more (eighteen rotations in all). The logic behind this sequence is that the performance of $M_D.B_L$ four times reorients the top front edge, which we want, and three edges elsewhere, which we don't; so we bring two front edges in turn to the top, reorient both, and let the unwanted reorientations elsewhere cancel themselves out. If we replace each of the operations $F_C$ and $F_A$ by $F_H$, we reorient the top and bottom edges instead of the top and left.

Of all the sequences which have been discovered, these seem to me to be the easiest to learn and remember. Nevertheless, they are far

Visible faces                    Hidden faces

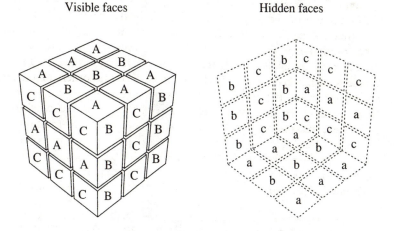

**Figure 7.20** Rubik's cube: an impossible configuration

from solving the cube with the smallest possible number of rotations, and experts deploy a battery of more sophisticated sequences. For details, see *Rubik's cubic compendium*.

Having got your cube into order, you may wish to create systematic decorative patterns on its faces. No new principle is involved, but you may encounter the same restriction that affects the fifteen puzzle: only even permutations are possible. For example, the attractive pattern shown in Figure 7.20 is unfortunately unattainable; five of the six faces can be set up as required, but the sixth cannot. The reason is that the permutation which forms Figure 7.20 from a correctly arranged cube consists of two three-cycles on corners and two three-cycles and a six-cycle on edges. A cycle of even length is an odd permutation, and a cycle of odd length is an even permutation, so the overall permutation is odd and the problem is insoluble.

This brings us back to the observation with which we started this chapter, that parity is the most important single weapon in the analysis of puzzles. So let us return to the chess board (Figure 7.21) and give the last word to the poet.

> Said the actress to the bishop, with a smile
> That was certainly intended to beguile,
> 'There are many things to do
> Which are much more fun for two;
> Won't you come on to my *black* square for a while?'

**Figure 7.21** The actress and the bishop

# 8

## SAUCE FOR THE GANDER

We now consider 'impartial' games of strategy: games in which the same moves are available to each player. The best known game in this class is nim, in which each player in turn removes objects from one of several piles. Indeed, it turns out that nim is not only one of the most widely played games of this type; it is in a sense the most fundamental, since all impartial games whose rules guarantee termination are equivalent to it.

## A winning strategy at nim

The rules of nim are very simple. The players divide a number of counters or other objects into piles, and each player in his turn may remove any number of counters from any one pile. The objective of play varies. In 'common' nim, which is the version usually played, a player's objective is to force his opponent to take the last counter. In 'straight' nim, his objective is to take it himself.

Of the two forms, straight nim is slightly the easier to analyse. It has a simple winning strategy. We imagine each pile divided into subpiles each of which contains an exact power of two: one, two, four, eight, and so on. This is always possible, and the division is always unique. Figure 8.1 shows piles of ten, nine, and six divided in this way. We now call a position *balanced* if it has an even number of subpiles of every size, and *unbalanced* if it contains an odd number of subpiles of any size. The position in Figure 8.1 is therefore unbalanced, since it has an odd number of subpiles containing one or four counters. If the pile of six is reduced to three, it becomes balanced (Figure 8.2).

Now if a position is unbalanced, it can always be balanced by reducing a single pile. The procedure is to locate the pile containing the largest unbalanced subpile (if there are several such piles, any of them may be chosen), temporarily to ignore this pile, to note any

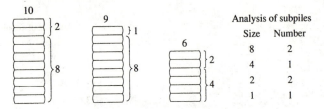

**Figure 8.1** Nim: an unbalanced position

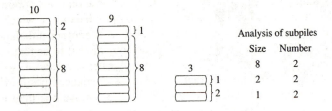

**Figure 8.2** Nim: a balanced position

unbalanced subpiles that now occur in the rest of the position, and to reduce the chosen pile to whatever is necessary to balance them. Thus in Figure 8.1, we observe that the size of the largest unbalanced subpile is four, and that this subpile occurs in pile 3; so we temporarily ignore this pile, note that the rest of the position now contains unbalanced subpiles of two and one, and reduce pile 3 to three counters in order to balance them. Another case is shown in Figure 8.3. Here, the size of the largest unbalanced subpile is eight, and it appears in pile 1; so we temporarily ignore this pile, note that the rest of the position now contains no unbalanced subpiles at all, and so remove pile 1 altogether.

On the other hand, if a position is already balanced, any reduction of a single pile unbalances it. We can see this by looking back at Figure 8.2. Let us consider a hypothetical reduction of pile 1. If we look at the rest of the position, we find that it contains odd numbers of subpiles of sizes eight and two. In order to balance these, pile 1 must consist of precisely these subpiles; no more, no less. But pile 1 already does consist of precisely these subpiles, since the position is already balanced, so any reduction must leave at least one subpile unbalanced. The same is true of a reduction of any other pile.

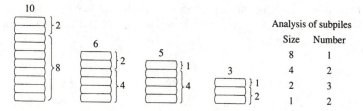

**Figure 8.3** Nim: another unbalanced position

A winning strategy now becomes clear. Suppose that our opponent has left us an unbalanced position. We promptly balance it, which we can always do, and thus we force him to give us another unbalanced position. This continues until there are no counters left to be removed. The last position of all (no counters in any pile) is balanced, so we win. Alternatively, suppose that he leaves us a balanced position. In theory, we have now lost; our best hope is to remove one man at a time, hoping that he will make a mistake, but we have no chance if he knows the game.

Having analysed straight nim, we can deal quickly with common nim. It might seem that common nim should be the reverse of straight nim; a winning position at straight nim should be a losing position at common nim, and vice versa. If every pile contains one counter only, this is indeed so. But if any pile contains more than one counter, the opposite is true; such a position is a winning position at common nim if and only if it is a winning position at straight nim. The winning strategy for common nim is in fact as follows.

(a) Play as at straight nim until your opponent presents you with a position in which only one pile contains more than one counter.

(b) At this point, play to leave an *odd* number of single-counter piles. (At straight nim, you would leave an even number of such piles.)

The theory of nim was first published by C. L. Bouton in 1901. It is one of the best of mathematical games: bewildering when you do not know how to play, simple and elegant when you do.

# Nim in disguise

Many games, when examined, prove to be simply disguised versions of nim. The disguise is usually of straight nim, in which a player's objective is to take the last counter himself and leave his opponent without a move. So, from now onwards, the unqualified word 'nim' means straight nim.

## (a) Northcott's game

This game is played on a chessboard or other similar board (Figure 8.4). Each player has a man on every file (vertical column), and he is allowed to move it any number of squares up or down the file. He may not jump over his opponent's man, nor move a man to another file, nor move two men at once. The winner is the player who makes the last move.

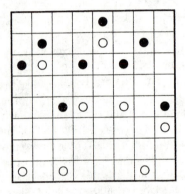

**Figure 8.4** Northcott's game

A little thought shows that the numbers of squares between the men behave like piles of counters at nim, except that both increases and decreases are allowed. But the increases prove to make no difference. The winner need never make one, because he can win without; and if the loser makes one, the winner can immediately move so as to restore the previous gap. It would be a different matter if the loser were allowed to make increasing moves indefinitely, but the edge of the board stops him. So Northcott's game is really a very simple disguise for nim. The winning move in Figure 8.4 is to reduce the gap in column 3 to one square.

## (b) The concertina game

This game is played on a continuous circuit which has been bent into

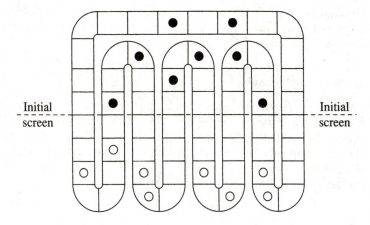

**Figure 8.5** The concertina game

a concertina (Figure 8.5). The board is placed so that half the U-bends are in front of each player, one player getting four small bends and the other three small bends and the large one. Each player then places two men on each of the bends in his half of the board, and play proceeds as in Northcott's game. The winning move in Figure 8.5 is to close the gap of two squares in the rightmost column but one.

To set up the initial position, the players place a screen across the middle of the board so that neither can see what the other is doing. This brings the game within the ambit of Chapter 6, and the task of determining an optimal initialization strategy for each player is drawn to the attention of any reader with a sufficiently large computer and nothing better to do. Once each player has chosen his position, the screen is removed. The rule determining first move depends on the precise size of the board. If the board is of the size shown in Figure 8.5, the player with the large bend must play second, since he can ensure an unbalanced position (and hence a win, if he is allowed to start) by placing his men as in Figure 8.6. The two men at one end of the large bend ensure that one gap consists of at least eight squares, and the remaining men ensure that the other gaps are smaller.

## (c) The silver dollar game

This game is played with a set of coins and a semi-infinite row of squares (Figure 8.7). Some of the squares are empty, some contain

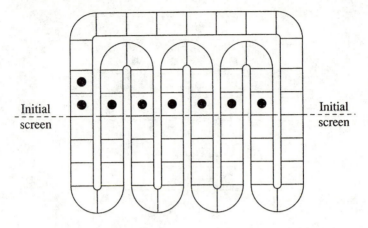

**Figure 8.6** Why the lower player must be allowed to start

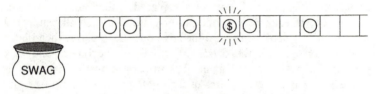

**Figure 8.7** The silver dollar game

worthless coins, and one contains a silver dollar. A player, on his turn, may move any one coin any number of unoccupied squares to the left, or pocket the leftmost coin. The winner is he who pockets the dollar. Conway credits the game to N. G. de Bruijn; he quotes it in a slightly more complicated form, but the differences are unimportant.

The key coin turns out to be not the dollar itself but the coin immediately to its left. We *don't* want to pocket this coin, because this will allow our opponent to pocket the dollar. So this coin may be regarded as a hot potato, and it is denoted by '!' in subsequent figures.

There are now two phases to the play: when there is at least one coin to the left of the hot potato, and when there is none. It is convenient to deal with the latter case first, although it is the later phase in actual play. Since nobody wants to pocket the hot potato, this phase effectively finishes when all the coins are jammed up against the extreme left-hand end of the row.

To see the relation of this phase to nim, we start at the rightmost coin and mark off the coins in pairs, noting the number of empty squares between the members of each pair. If the number of coins is odd, we also note the number of empty squares to the left of the last coin (which we call the 'widow', and which at this stage is the hot potato). Figure 8.8 shows the two cases, Figure 8.8(a) showing a position with a widow and Figure 8.8(b) a position without. It is now easily seen that the numbers of squares behave like the gaps in Northcott's game, and hence like nim piles. The good move is arrowed in each case. The position in Figure 8.8(b) can also be won by increasing the rightmost gap from 0 to 2, but the winner never needs to increase a gap in this way and the analysis is simplified if we ignore the possibility.

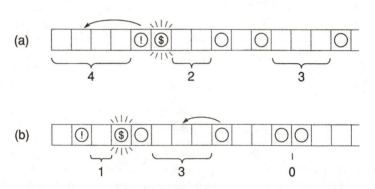

**Figure 8.8** The silver dollar game: the second phase

The analysis of the first phase is similar, except that if we have a widow, we now add one to the number of empty squares to its left (because we are now willing to pocket this coin, so the pocket itself behaves like an empty square). Figure 8.9 shows the two cases, with and without a widow, and the good moves are again arrowed. Note that there are three equally good winning moves in Figure 8.9(a).

The transition from one phase to the other appears to require special treatment, but on closer inspection this proves not really to be the case. The winner never needs to pocket the left-hand member of a pair, and he can always move the new widow to an appropriate

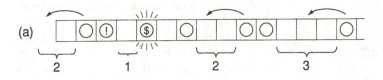

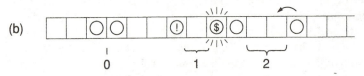

**Figure 8.9** The silver dollar game: the first phase

position if the loser does so. The winner pockets a widow only in order to leave a balanced position in the reduced game, as in Figure 8.9(a); and if the loser pockets a widow, he leaves an unbalanced position in the reduced game, and the winner can balance it as usual.

So the silver dollar game is yet another disguise for nim.[1]

# All cul-de-sacs lead to nim

The essential properties of nim are these: (i) every pile has a measure; (ii) if this measure is non-zero, it is possible to move to a pile with any given smaller measure; (iii) it is not possible to move to another pile with the same measure. We shall now show how similar measures can be assigned to the positions of any impartial game whose rules guarantee termination. This fundamental property, first discovered by R. P. Sprague and independently rediscovered by P. M. Grundy, proves that any such game is equivalent to nim.

The question of guaranteed termination deserves a brief examination. Nim and the silver dollar game are guaranteed to terminate however well or badly they are played, nim because every move decreases the number of counters remaining, the silver dollar game because every move takes 'a coin towards a dead end. We may call such games 'cul-de-sacs'. Northcott's game and the concertina game are not cul-de-sacs; there is nothing in their rules to stop the players from

---

[1] Readers who teach computing science in schools might like to consider the silver dollar game as an exhibition project. Its simple rules and linear board are ideally suited to play by an elementary computer-controlled robot.

moving back and forth indefinitely. It is true that the winner can ensure termination even though the rules do not, and a generalized version of the theory can be shown to apply to these games; but the exposition is more complicated, and we shall not pursue it.

Now if a game is a cul-de-sac, its positions can be divided into sets according to the greatest number of moves that the game may still take. Thus we can define set $\{P_0\}$ to comprise the positions in which play has already terminated, set $\{P_1\}$ to comprise the positions in which play is guaranteed to terminate after one more move, and so on. Thus a position in set $\{P_n\}$ has an immediate successor in set $\{P_{n-1}\}$, and may have others scattered among sets $\{P_0\}$ to $\{P_{n-1}\}$ inclusive.

We can now take the sets in order, and assign a value to every position. The detailed procedure is as follows.

(a) To each position in set $\{P_0\}$, we assign the value 0.

(b) To each position in set $\{P_1\}$, we assign the value 1.

(c) Set $\{P_2\}$ is more complicated. Some of its positions may have immediate successors only in $\{P_1\}$, others may have them in both $\{P_1\}$ and $\{P_0\}$. The rule is now simple: we assign to each position *the lowest value not assigned to any of its immediate successors*. If a position has immediate successors in both $\{P_1\}$ and $\{P_0\}$, we assign it the value 2. If it has immediate successors only in $\{P_1\}$, it has no immediate successor with value 0, so we assign it this value.

And so it goes on. At each stage, we assign to every position the lowest value not assigned to any of its immediate successors. We call the value so assigned the *Grundy value* of a position. (It should perhaps be called the 'Sprague value' on the grounds of priority, but Sprague's work was unknown in this country for many years and 'Grundy value' has long been accepted usage in English; and Grundy, apparently unlike Sprague, published follow-up work.)

The crucial properties of Grundy values follow immediately from the way in which they are assigned. No immediate successor of a position has the same Grundy value as the position itself; so, in particular, the effect of any move from a position with Grundy value 0 (if a move is possible at all) is to produce a position whose Grundy value is not 0. Conversely, if a position has a non-zero Grundy value $g$, and $h$ is any number lower than $g$, it is possible to move to a position with Grundy value $h$.

It is now clear that such a game, if played on its own, behaves like nim. A player who is presented with a position with a non-zero Grundy value can always move to a position with Grundy value 0; his opponent, if he can move at all, must then move to another position with a non-zero Grundy value; and the process repeats. What is more important, however, is that a *composite* of such games also behaves like nim. Suppose that we have such a composite, and that the Grundy values of the constituents are $a$, $b$, ..., $m$. If we have nim piles of sizes $a$, $b$, ..., $m$, and the composite position is unbalanced, there is always a reduction of one pile that will balance it. So it is here; there is always a Grundy value which can be reduced so as to balance the composite position, and there is always a move in the constituent game that will yield the required new value. Conversely, if the composite position is already balanced, any move in a constituent game changes one of the constituent Grundy values, and so unbalances the composite.

# Grundy analysis in practice

To put some flesh on this theory, let us assign Grundy values to some simple games.

### (a) *Take one, two, or three*

In the days of my youth, we had a playground game in which each player removed either one, two, or three matches from a pile. The objective was to make the opponent remove the last match, but for the moment let us analyse the corresponding game in which a player seeks to remove the last match himself.

Let us denote a pile of $n$ matches by $M_n$. The assignment now proceeds as follows.

(i) No play is possible from the notional pile containing no matches, $M_0$, so its Grundy value is 0.

(ii) From $M_1$, play is possible only to $M_0$. This has Grundy value 0, so the Grundy value of $M_1$ is 1.

(iii) From $M_2$, play is possible to $M_0$ or $M_1$. These have Grundy values 0 and 1 respectively, so the Grundy value of $M_2$ is 2.

(iv) From $M_3$, play is possible only to $M_0$, $M_1$, or $M_2$. These have Grundy values 0, 1, and 2, so the Grundy value of $M_3$ is 3.

(v) At $M_4$, we have a change. Play is now possible to $M_1$, $M_2$, or

$M_3$, but not to $M_0$. The attainable positions have Grundy values 1, 2, and 3, and the lowest number not included in this list is 0. So the Grundy value of $M_4$ is 0.

(vi) Similarly, from $M_5$, play is possible only to $M_2$, $M_3$, or $M_4$. These positions have Grundy values 2, 3, and 0, so the Grundy value of $M_5$ is 1.

We can pursue this calculation a little further, but it is becoming clear that we have a repetitive pattern: if the division of $n$ by 4 leaves remainder $j$, the Grundy value of $M_n$ is $j$. It is not uncommon for a Grundy sequence to become periodic, either from the start (as here) or at some later stage. In practical terms, the winning strategy in this particular game is always to leave a multiple of four. This could have been worked out easily enough without the aid of the general theory, but harder games are about to follow.

### (b) Kayles (*Rip van Winkle's game*)

This game, which dates back to Sam Loyd, has various guises. Perhaps the simplest uses a miniature cannon and a row of hinged targets (Figure 8.10). The cannon is assumed to fire a ball of fixed size, and the targets to be spaced so that a shot can knock down either a single target or two adjacent ones. The cannon is assumed to be reliable and the players to be perfect shots, and shooting to miss is forbidden. Each player takes one shot in turn, and the object is to be the player who knocks down the last target.

If the row has no gaps, the first player has an easy win. If the number of targets in the row is odd, he knocks down the middle one; if it is even, he knocks down the middle pair. This presents his opponent with two separate rows of equal size, and whatever his opponent now does in one row, he can echo in the other.

If the row has gaps, however, the game is by no means trivial. A row with gaps can be regarded as a composite game, each constituent comprising a row without gaps, so let us denote a row of $n$ adjacent targets by $K_n$, and let us use '$+$' to denote composition (in other words, '$K_m + K_n$' will denote the composite game whose constituents are $K_m$ and $K_n$). The Grundy value of $K_0$ is 0, because we cannot move at all; that of $K_1$ is 1, because we can move only to $K_0$; that of $K_2$ is 2, because we can move either to $K_0$ or to $K_1$. This brings us to $K_3$, from which we have three moves: to $K_1$, to $K_2$, or to $K_1 + K_1$.

We can deal with $K_1 + K_1$ in two ways. We can analyse it from first

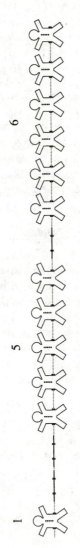

**Figure 8.10** Kayles (Rip van Winkle's game)

principles: from $K_1 + K_1$, we can move only to $K_1$; the Grundy value of $K_1$ is 1; therefore that of $K_1 + K_1$ is 0. Alternatively, we can invoke the Sprague–Grundy theorem, and use the fact that the Grundy value of a composite game is the number which balances the Grundy values of the constituents. The number which balances 1 and 1 is 0, so the Grundy value of $K_1 + K_1$ is 0. This is much the more powerful technique, since it means that we need to perform a full analysis only on complete rows; everything else can be done by calculating balances.

Either way, we find that the Grundy value of $K_3$ is 3, since this is the lowest number not contained in the set 1 ($K_1$), 2 ($K_2$), and 0 ($K_1 + K_1$).

Subsequent values can be calculated similarly; for example, the Grundy value of $K_4$ is 1, since this is the lowest number not contained in the set 2 ($K_2$), 3 ($K_3$), 0 ($K_1 + K_1$), and 3 again ($K_1 + K_2$). The sequence up to $K_{20}$ is shown in the first row of Table 8.1, and the complete sequence, which proves to have period 12 from $K_{71}$ onwards, appears both in *On numbers and games* by J. H. Conway (Academic Press, 1976) and in *Winning ways for your mathematical plays* by E. R. Berlekamp, Conway, and R. K. Guy, hereinafter 'BCG' (Academic Press, 1982). The periodicity was first proved by Richard Guy in 1949, the sequence necessarily being calculated by hand. Nowadays, such a calculation provides a simple exercise on a home computer, and some readers may be tempted to repeat it.

The remaining rows of Table 8.1 show the Grundy values of the games which we can reach from $K_n$. If we knock down one man, we obtain one of the values in the column immediately underneath; if we

**Table 8.1** Kayles: the first twenty Grundy values

| $n$ | 0 | 1 | 2 | 3 | 4 | 5 | 6 | 7 | 8 | 9 | 10 | 11 | 12 | 13 | 14 | 15 | 16 | 17 | 18 | 19 | 20 |
|---|---|---|---|---|---|---|---|---|---|---|---|---|---|---|---|---|---|---|---|---|---|
| $K_n$ | 0 | 1 | 2 | 3 | 1 | 4 | 3 | 2 | 1 | 4 | 2 | 6 | 4 | 1 | 2 | 7 | 1 | 4 | 3 | 2 | 1 |
| $K_{n-1}$ | | 0 | 1 | 2 | 3 | 1 | 4 | 3 | 2 | 1 | 4 | 2 | 6 | 4 | 1 | 2 | 7 | 1 | 4 | 3 | 2 |
| $K_{n-2}+K_1$ | | | | 0 | 3 | 2 | 0 | 5 | 2 | 3 | 0 | 5 | 3 | 7 | 5 | 0 | 3 | 6 | 0 | 5 | 2 |
| $K_{n-3}+K_2$ | | | | | | 0 | 1 | 3 | 6 | 1 | 0 | 3 | 6 | 0 | 4 | 6 | 3 | 0 | 5 | 3 | 6 |
| $K_{n-4}+K_3$ | | | | | | | | 0 | 2 | 7 | 0 | 1 | 2 | 7 | 1 | 5 | 7 | 2 | 1 | 4 | 2 |
| $K_{n-5}+K_4$ | | | | | | | | | | 0 | 5 | 2 | 3 | 0 | 5 | 3 | 7 | 5 | 0 | 3 | 6 |
| $K_{n-6}+K_5$ | | | | | | | | | | | | 0 | 7 | 6 | 5 | 0 | 6 | 2 | 0 | 5 | 6 |
| $K_{n-7}+K_6$ | | | | | | | | | | | | | | 0 | 1 | 2 | 7 | 1 | 5 | 7 | 2 |
| $K_{n-8}+K_7$ | | | | | | | | | | | | | | | | 0 | 3 | 6 | 0 | 4 | 6 |
| $K_{n-9}+K_8$ | | | | | | | | | | | | | | | | | | 0 | 5 | 3 | 7 |
| $K_{n-10}+K_9$ | | | | | | | | | | | | | | | | | | | | 0 | 6 |

knock down two men, we obtain one of the values in the column immediately to its left. These values allow the winning move in any position to be found quickly and easily. For example, Figure 8.10 shows rows of one, five, and six men, and the Grundy values of $K_1$, $K_5$, and $K_6$ are 1, 4, and 3 respectively. Now if we had nim piles of sizes 1, 4, and 3, the winning move would be to reduce the pile of size 4 to 2, so the requirement in Figure 8.10 is to replace $K_5$ by something whose Grundy value is 2. The relevant columns of Table 8.1 show that the only candidate is $K_3 + K_1$, so the winning move is to knock down a single man one from the end of the row of five.

### (c) Dots and loops

There is an interesting class of games in which the permitted move is to draw a line through one or more dots, no line being allowed to cross an existing line. Many of these games are equivalent to games with counters; for example, if we have $n$ dots and we draw a loop through $m$ of them, enclosing $i$ dots and leaving $j$ outside, it is as if we had removed $m$ counters from the middle of a row of $n$, leaving separate rows of lengths $i$ and $j$. So if we allow $m$ to be 1 or 2 then we get a game equivalent to kayles, and if we allow $m$ to be any positive number then we get yet another simple disguise for nim. But it is easier to draw dots on paper than to set up rows of counters, and BCG comment that many people appear to find the games more attractive in this form.

The most obvious rule is to draw a loop through a single dot, but the resulting game proves to be trivial. Whatever the previous play, a loop can always be drawn through any dot still untouched, so the first or second player has an automatic win according as the number of dots is odd or even. But if we require the loop to pass through precisely *two* dots (Figure 8.11), we obtain a game which may appropriately be called 'Dawson's loops' in honour of T. R. Dawson,[2]

[2] This name is rooted in history. Dawson is remembered primarily as an outstanding composer of generalized chess problems, but he also took a strong interest in other mathematical matters, and many of his six thousand 'chess' problems had a pronounced mathematical flavour. One was a simple but elegant pawn problem which stimulated the interest of Guy. This problem can be shown to be mathematically equivalent to the drawing of loops through two dots and to various other games, and it has become customary, following Guy, to attach Dawson's name to all these games. There is an element of homage in this, since only the pawn problem was apparently considered by Dawson himself and the ultimate analysis even of this problem appears to owe as much to Guy as to Dawson; but it is homage in which I am happy to join.

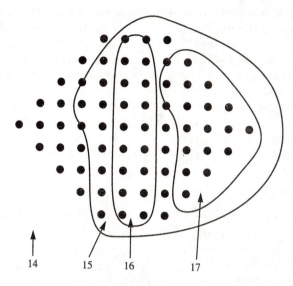

**Figure 8.11** Dawson's loops

and this game has quite a different behaviour. If the number of dots *n* is even, the first player has an easy win, since he can draw his loop so as to divide the remaining dots evenly and then echo his opponent's moves; but if *n* is odd, the result is not obvious, and a full Grundy analysis is necessary to determine the winner. In fact the second player wins if *n* is of the form $34k + l$ where $l = 5, 9, 21, 25,$ or 29, and also if $n = 1, 15,$ or 35; otherwise, the first player wins. The first twenty values of the Grundy sequence are shown in the first row of Table 8.2, and the complete sequence, which Guy proved in 1949 to have period 34 from $D_{53}$ onwards, appears in *Winning ways for your mathematical plays*. Again, the calculation can easily be repeated using a home computer.

The remainder of Table 8.2 shows the Grundy values of the games which we can reach from $D_n$, and it is even simpler to use than Table 8.1 because we need look only at the column immediately below the game of interest. For example, Figure 8.11 shows regions containing fourteen, fifteen, sixteen, and seventeen dots, and the Grundy values of $D_{14}$, $D_{15}$, $D_{16}$, and $D_{17}$ are 4, 0, 5, and 2 respectively. So the requirement is to replace $D_{17}$ by something whose Grundy value is 1, and Table 8.2 shows that the only candidate is $D_{11} + D_4$. Hence the winning move is to draw a loop within the region of seventeen dots,

**Table 8.2** Dawson's loops: the first twenty Grundy values

| $n$ | 0 | 1 | 2 | 3 | 4 | 5 | 6 | 7 | 8 | 9 | 10 | 11 | 12 | 13 | 14 | 15 | 16 | 17 | 18 | 19 | 20 |
|---|---|---|---|---|---|---|---|---|---|---|---|---|---|---|---|---|---|---|---|---|---|
| $D_n$ | 0 | 0 | 1 | 1 | 2 | 0 | 3 | 1 | 1 | 0 | 3 | 3 | 2 | 2 | 4 | 0 | 5 | 2 | 2 | 3 | 3 |
| $D_{n-2}$ | | | 0 | 0 | 1 | 1 | 2 | 0 | 3 | 1 | 1 | 0 | 3 | 3 | 2 | 2 | 4 | 0 | 5 | 2 | 2 |
| $D_{n-3}+D_1$ | | | | 0 | 1 | 1 | 2 | 0 | 3 | 1 | 1 | 0 | 3 | 3 | 2 | 2 | 4 | 0 | 5 | 2 | |
| $D_{n-4}+D_2$ | | | | | 0 | 0 | 3 | 1 | 2 | 0 | 0 | 1 | 2 | 2 | 3 | 3 | 5 | 1 | 4 | | |
| $D_{n-5}+D_3$ | | | | | | 0 | 3 | 1 | 2 | 0 | 0 | 1 | 2 | 2 | 3 | 3 | 5 | 1 | | | |
| $D_{n-6}+D_4$ | | | | | | | 0 | 2 | 1 | 3 | 3 | 2 | 1 | 1 | 0 | 0 | 6 | | | | |
| $D_{n-7}+D_5$ | | | | | | | | | | 0 | 3 | 1 | 1 | 0 | 3 | 3 | 2 | 2 | | | |
| $D_{n-8}+D_6$ | | | | | | | | | | | | 0 | 2 | 2 | 3 | 0 | 0 | 1 | | | |
| $D_{n-9}+D_7$ | | | | | | | | | | | | | | 0 | 0 | 1 | 2 | 2 | | | |
| $D_{n-10}+D_8$ | | | | | | | | | | | | | | | | 0 | 1 | 2 | | | |
| $D_{n-11}+D_9$ | | | | | | | | | | | | | | | | | | 0 | | | |

leaving eleven dots inside and four outside (or vice versa). It is an instructive exercise to verify that any other move allows the opponent to win.

# Some more balancing acts

The technique of balanced subdivision (in other words, of dividing a game into two parts such that a player can now echo in one part whatever his opponent may do in the other) provides a powerful weapon in the play of strategic games. As a brief digression from the main theme of this chapter, let us look at a few more examples.

## (a) Division into three

The move in this game is to divide a pile of at least three counters into three smaller piles. A single-pile game is not quite trivial, but balanced subdivision shows it to be a straightforward win for the first player. He makes one pile of one or two counters, depending on whether the initial pile is odd or even, and divides the rest into two piles of equal size. The pile of one or two is now immovable, and he can echo any move that his opponent makes in the equal piles.

A similar technique can be applied to any game requiring a pile to be divided into more than two parts. Requiring a division into $n$ parts, the first player divides the initial pile into $(n-3)$ piles each containing one counter, one pile containing one or two counters, and two equal piles.

## (b) Mother's nightmare

A game popular among small children[3] is to raid newly baked cakes as soon as the cook's back is turned. We can formalize this game by arranging the cakes in rectangles, and allowing each player in turn to remove a complete line (row or column) from a single rectangle. If the line is taken from the middle of a rectangle, the two new rectangles are subsequently treated as separate. The traditional objective is to scoff as many cakes as possible, but for present purposes let us assume that our objective is to take the last cake and leave our opponent without a move.

Let us start by supposing that we have a single rectangle of size $m \times n$. If either $m$ or $n$ is odd (Figure 8.12, left), the first player has an easy win; he removes the middle line and leaves a balanced pair (unless the rectangle is only one line deep, in which case he removes the whole of it and wins immediately). But if $m$ and $n$ are both even (Figure 8.12, right), it is the second player who wins, since he can mirror his opponent's move. Anything left outside the two lines which have been removed now forms a balanced pair, and anything left between them forms a smaller rectangle which again has even sides.

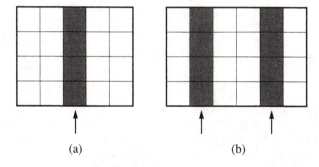

(a)                    (b)

**Figure 8.12** Having a cake and eating it

A position containing more than one rectangle can be analysed by classifying its constituents as 'even–even', 'odd–odd', and 'odd–even' according to the parities of their sides. The even–even rectangles are second-player wins, and can be ignored (because the winner needed never play in such a rectangle on his own initiative, and can make the appropriate winning reply if his opponent does so). This leaves

---

[3] And husbands. S.B.

odd–odd and odd–even rectangles, and we now observe that a pair of odd–odd rectangles can be treated as a balanced pair even if the actual sides are unequal, and that a pair of odd–even rectangles can be treated similarly. For example, the moves that may be possible in an odd–odd rectangle fall into four classes (Figure 8.13): (i) reduction to a single odd–even rectangle; (ii) division into two odd–even rectangles; (iii) division into two odd–odd rectangles; and (iv) removal of the entire rectangle. A move in class (i) can be answered by a further reduction to another odd–odd rectangle, restoring the previous balance; and the effect of a move in any of classes (ii) to (iv) is to replace a single rectangle by a balanced pair (or to remove it altogether), and the overall balance can be restored by making another such move in another odd–odd rectangle. The analysis of an odd–even pair is similar.

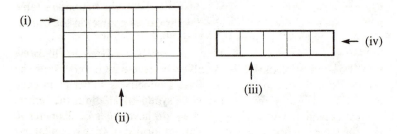

**Figure 8.13** The four classes of operation on an odd–odd rectangle

So if a position contains an even number of odd–odd rectangles, and another even number of odd–even rectangles, it is a second-player win. The sizes of the odd–odd and odd–even rectangles are irrelevant; all that matters is that there be an even number of each. On the other hand, if there is an odd number of rectangles in either category, the first player can win, since a move is always available to him which leaves even numbers of rectangles in each category. The same conclusion can be reached by a normal Grundy analysis, since such an analysis can be shown to assign 0 to every even–even rectangle, 1 to every odd–odd, and 2 to every odd–even; but the analysis in terms of balanced pairs is more instructive.

### (c) Rich man's table, poor man's table

This game shows how balanced subdivision can be used even when the playing space is continuous and the number of alternative moves is infinite. The story starts when a rich man visits a poor man and inveigles him into a gamble in which the players alternately place coins on a table, the first person to be unable to place a coin losing. The poor man, as the host, courteously offers his opponent the first move; but he has a plain rectangular table, and the rich man starts by placing a coin exactly at its centre. He can now match the poor man's every coin, and place his own in a diametrically opposite position. So the poor man runs out of moves first, and loses all his money.

There is a sequel. The poor man scrimps and saves, and works long hours at the pitiful wages that the rich man is prepared to pay; and eventually he accumulates enough to challenge the rich man to a return match. He calculates, correctly, that the rich man will offer hospitality this time and will concede the first move to his guest; but when they actually start to play, he discovers that the rich man's table has an ornate silver candlestick (symmetrical, of course) in the centre. So the poor man is fleeced once more.

Certain generalizations of this game can be analysed in the same way. For example, the table need not be rectangular, nor need the central obstacle be circular; diametral symmetry is sufficient in each case. But an obstacle must not only be symmetrical about the centre, it must actually cover it. Suppose that we have coins of diameter $d$, that the table is in the shape of a diamond formed from two equilateral triangles of side $d$, and that the obstacle consists of two protruding nails a distance $d\sqrt{3}/2$ apart. No general statement can now be made, since the winner depends on the orientation of the obstacle. If it is along the long axis (Figure 8.14, left), the first player wins; he places his coin as shown, and the second player has nowhere to go. On the other hand, if the obstacle is along the short axis (Figure 8.14, right),

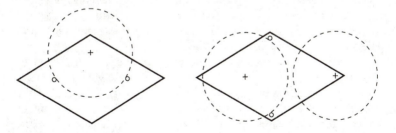

**Figure 8.14** First player's table, second player's table

the second player wins; each player has room for just one coin in his own half, and neither can encroach sufficiently far across the centre to inconvenience the other.

# Playing to lose

At the start of this chapter, it was observed that nim is usually played with the rule that the last player to move *loses*, and the same is true of 'take one, two, or three' in the form in which I encountered it at school. Such play is known as 'misère' play. The general theory of misère play is markedly less satisfactory than that of normal play, but we must have a brief look at it. We continue to assume that the game is a cul-de-sac.

In an earlier section, we found a procedure for assigning a Grundy value to every position in a game. It is possible to assign a 'misère Grundy value' similarly. The procedure is to assign 1 to all positions in the set $\{P_0\}$ and 0 to all positions in the set $\{P_1\}$, and then to continue as before. Thus if we analyse the misère form of 'take one, two, or three', we find ourselves assigning 1 to $M_0$, 0 to $M_1$, 2 to $M_2$, 3 to $M_3$, and the values now repeat: 1 to $M_4$, 0 to $M_5$, and so on. The winning strategy is to leave a position with value 0, which here means playing to leave one more than a multiple of four.

This really seems rather simple; why the gloom? The trouble is that the misère Grundy value merely tells us who wins if the game is played on its own; *it does not tell us what happens if the game is played as part of a composite*. There is no general analogue of the Sprague–Grundy theorem for misère play; the strategy which we used for common nim, of playing normally until just before the end, works for some games, but it does not work for all. In consequence, the work required to calculate a misère Grundy sequence may be overwhelming. When we calculate a normal Grundy sequence, we can deal with occurrences of composite games by calculating balancing values. No such simplification is available for general misère play.

For example, in normal kayles, the calculation of the Grundy value of $K_n$ involves consideration of the $2n-5$ composite games which are immediate successors of $K_n$, but each of these composites requires the calculation only of a single balancing value and so the amount of work required increases only linearly with $n$. The calculation of the misère Grundy value requires consideration of the same $2n-5$ composite games, but some of them must now be analysed in depth, and the amount of work required appears to increase rather more than exponentially with $n$.

Even some of the most basic properties of normal games (for example, that a balanced pair is a win for the second player and that the addition of a balanced pair does not change the result of a game) are not always true of misère play. In misère kayles, the game $K_4$ is easily seen to be a second-player win; if the first player knocks down one man, the second knocks down two, and vice versa. Similarly, the balanced pair $K_2 + K_2$ is a second-player win. Yet the composite games $K_4 + K_4$ and $K_4 + K_2 + K_2$ (Figure 8.15) are first-player wins, since the first player can knock out the two middle men from $K_4$ and leave $K_1 + K_1 + K_4$ or $K_1 + K_1 + K_2 + K_2$ with his opponent to play. So a balanced pair may be a first-player win even though its constituents are second-player wins, and the addition of a balanced pair may change the result of a game even though the balanced pair is a second-player win when played on its own.

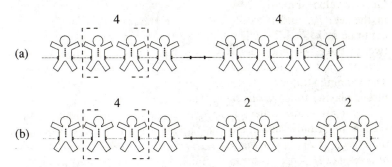

**Figure 8.15** Misère kayles: two surprising composites

Misère play is among the topics discussed by Conway in *On numbers and games* and by BCG in *Winning ways for your mathematical plays*, and you should consult those books if you want to know more. But the subject is incomplete, and much of the theory that exists is difficult. The fact that games such as nim and 'take one, two, or three' are usually played in misère form reflects no more than an instinctive feeling that it is more elegant for a player to compel his opponent to do something than to do it himself; but the absence of a satisfactory general theory of misère play provides an interesting comment on this instinct.

# 9

# THE MEASURE OF A GAME

We now come to one of the most interesting of modern developments: the assessment of the inherent measure of a game. This is the simplest case of the theory of 'numbers and games' which is due to Conway. Many games appear to have measures, of course, in that their scoring systems allot points for wins of various kinds, but these measures are essentially arbitrary. Our measures here are natural, and owe nothing to superimposed scoring systems.

## Nim with personal counters

To begin our investigation, let us consider a version of nim in which each of the counters is owned by one player or the other. The object of play remains as in ordinary straight nim, to leave the opponent without a move, but a player may remove a set of counters from a pile only if the *lowest* counter that he removes is one of his own. Anticipating some of our conclusions, we call the players Plus and Minus, and we label the counters accordingly. Thus in Figure 9.1(a), Plus can remove either the top counter or the top three, Minus either the top two or the whole pile; while in Figure 9.1(b), Plus can remove any or all of the counters, whereas Minus cannot remove any at all.[1]

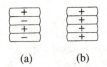

(a)          (b)

**Figure 9.1** Personalized nim piles

---

[1] I do not know who first conceived nim with personal counters, but as a game it is surely ancient. The recognition of its numerical properties appears to be due to Berlekamp.

In the simplest games of this kind, every counter in a pile belongs to the same player. Such games are almost trivial, but they provide a convenient demonstration of the basic rules. In particular, each pile has a measure which is simply the number of counters in it, and the result of a game can be found by adding the measures of the piles; Plus wins if the sum is positive, and Minus if it is negative. And if it is zero, as in Figure 9.2? Provided that both players play sensibly, the result depends on the move; each player removes only a single counter at a time, so whoever has to move first runs out of counters first, and the game is a win for his opponent. This last rule, that *a game of zero measure is a win for the second player*, is fundamental to the theory.

**Figure 9.2** A zero game

Some further basic rules are conveniently stated here. Corresponding to every game $X$ is a complementary game in which the ownership of every counter is reversed. Let us call this complementary game '$-X$'. The complementary game provides a way of comparing the magnitudes of two games $X$ and $Y$, because we can set up the composite game $X+(-Y)$ and see who wins. We say that

$$X > Y$$

if the composite game $X+(-Y)$ is a win for Plus irrespective of who starts,

$$X < Y$$

if it is a win for Minus irrespective of who starts, and

$$X = Y$$

if it is a win for the second player. Figure 9.3(a) shows the case where

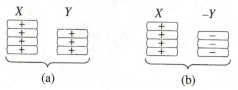

(a)                    (b)

**Figure 9.3** The comparison of games

$X$ contains four positive counters and $Y$ only three. Common sense suggests that $X > Y$, and Figure 9.3(b) demonstrates that our rules do indeed lead to this conclusion; the composite game $X + (-Y)$, shown in the figure, is easily seen to be a win for Plus irrespective of the start.

Common sense also suggests, trivially, that $X = X$, and the interpretation of this in accordance with our rules is that the composite game $X + (-X)$ should be a win for the second player. Figure 9.4(a) shows the case where $X$ is a pile of four positive counters, and we observe that the second player does indeed win. But the rule does not apply only to homogeneous piles. In Figure 9.4(b), the piles $X$ and $-X$ contain counters belonging to each player, but the composite game $X + (-X)$ is still a win for the second player; whatever the first player does in one component, his opponent can copy in the other. This is indeed a very general rule which does not apply only to personalized nim: *provided that the rules of the game ensure termination sooner or later*, a composite game of the form $X + (-X)$ is always a win for the second player.

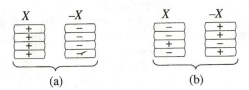

**Figure 9.4** Two balanced pairs

# Games of fractional measure

Although piles of homogeneous counters serve to illustrate the basic rules, they are not really very interesting. We now assume that at least one pile contains counters owned by both players.

The simplest such pile is that shown in Figure 9.5(a). Let us call this game $G$. Played on its own, it is clearly a positive game. Plus, to play, can remove the whole pile, leaving Minus without a move; Minus, to play, can remove only the upper counter, and Plus can still remove the lower. On the other hand, $G$ is outweighed by the game $P_1$, shown in Figure 9.5(b), which comprises just one positive counter. The complement of $P_1$ is the game which comprises one negative counter, and the composite game $G + (-P_1)$, shown in Figure 9.5(c),

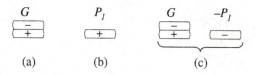

$$G \qquad\qquad P_1 \qquad\qquad G \qquad -P_1$$

(a)           (b)            (c)

**Figure 9.5** A pile of fractional measure

is easily seen to be a win for Minus. Plus, to play, can only remove the whole of $G$, leaving Minus to remove $-P_1$; Minus, to play, can remove just the top counter of $G$, leaving a balanced pair. So Minus wins the composite game $G+(-P_1)$ irrespective of who starts, and it follows that $G < P_1$.

So if the game $G$ has a measure which follows the normal rules of arithmetic, this measure must lie somewhere between 0 and 1. Perhaps guesswork suggests 1/2, but we need not resort to guesswork; we can test our hypotheses in a proper manner. If the measure of $G$ is indeed 1/2, it follows that the measure of $G+G$ must be 1, and hence that the composite game $G+G+(-P_1)$ must be a win for the second player. This game is shown in Figure 9.6, and the second player is indeed seen to have a win. The arrows in Figure 9.6(a) show best play with Plus starting; he can only start by removing one pile $G$, but Minus can reply by removing the top counter from the other pile $G$, and this creates a balanced pair which leaves Plus helpless. Similarly, Figure 9.6(b) shows best play with Minus starting; he can do no better than to remove the top counter from one of the piles $G$, but Plus can remove the other pile $G$ in reply, and again we have a balanced pair. So $G+G+(-P_1)$ is indeed a win for the second player, and it is reasonable to declare the measure of $G$ to be 1/2.

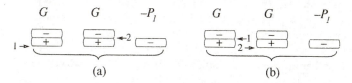

$$G \qquad\quad G \qquad -P_1 \qquad\qquad G \qquad\quad G \qquad -P_1$$

(a)                       (b)

**Figure 9.6** Fractional measure quantified

The next game to consider, perhaps, is the game $H$ shown in Figure 9.7(a). This is clearly positive though not as strongly positive as $G$, and we might perhaps conjecture that its measure is 1/4. The natural

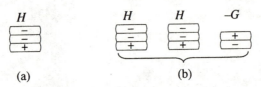

**Figure 9.7** A pile worth a quarter of a counter

way to test this conjecture is to play the game $H+H+(-G)$, which is shown in Figure 9.7(b). If Plus starts, his best move is to take a pile $H$, but Minus can reply by removing the top counter from the other pile $H$, and this leaves a balanced pair. Similarly, if Minus starts, his best move is to remove the top counter from one of the piles $H$, but Plus can reply by taking the other pile $H$. So $H+H+(-G)$ is a second player win, and it is reasonable to assign the measure 1/4 to $H$.

But does it follow, say, that $H+H+H+H$ is equal in measure to $P_1$? Yes, it does, as may be verified by setting up the composite game $H+H+H+H+(-P_1)$ and playing it out; but it also follows from a general rule of substitution. In ordinary arithmetic, if we have a true statement involving some quantity $A$ and we also know that $A=B$, we can substitute $B$ for $A$ and obtain another true statement. A similar substitution rule can be proved to apply to games, and it follows that we can substitute $H+H$ for $G$ in the statement $G+G=P_1$ and so derive the statement $H+H+H+H=P_1$.

It is now becoming clear that we can expect to find a game with any measure which is an exact binary fraction (a fraction whose denominator is a power of 2). We can obtain any required negative power of 2 (say $2^{-n}$) by placing $n$ negative counters on top of one positive counter, and we can obtain any other exact binary fraction by addition of piles. But can we find a *single* pile whose measure is any given binary fraction?

Yes, we can. From the ground up, each counter has an effective value of 1 until there is a change in ownership, after which the value of a counter is half of that immediately below it. A pile whose measure is any required finite binary fraction can now be obtained by the following simple procedure: if the measure of the pile so far is less than the target, we add a positive counter; if it is greater, we add a negative counter. Figure 9.8 shows this process applied to various numbers.

The assignment of values to individual counters allows us quickly

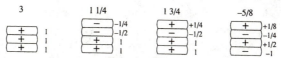

**Figure 9.8** Some measures embodied

to find the best move in games of this type. We glossed over this matter earlier in order not to clutter the discussion, but in fact the proof that a composite game such as $H + H + (-G)$ in Figure 9.7 is a win for the second player does require the second player to move sensibly. For example, if Minus starts and removes the top counter from the left-hand pile, Plus's *only* good reply is to remove the middle pile. To prove this from first principles is slightly tedious, but the values of the counters make it obvious. Figure 9.9 shows the position after Minus's first move. The sum of the counters in Figure 9.9 is 1/4, and if Plus removes the middle pile, the sum of the remaining counters will be 0, and the resulting game will be a zero game (and hence a win for Plus, since it will now be Minus's turn to play); but if Plus plays other than by removing the middle pile, the sum of the remaining counters will be −1/4, and the resulting game will be a win for Minus.

There remain only those numbers which are not finite binary fractions: 1/3, for example. No *finite* pile can have a measure which is such a number. However, it is possible that an infinite pile may have such a measure, and we shall return to this point later in the chapter.

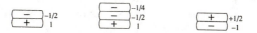

**Figure 9.9** How to find the only good move

# General piles

We now reintroduce neutral counters, and allow a player to remove one or more counters from a pile provided that the lowest counter removed is either one of his own counters or a neutral.

The simplest piles of this type are ordinary nim piles, which contain neutral counters alone. We shall henceforth call such piles 'neutral' piles. A single neutral pile is a win for the first player, and such a game is called 'fuzzy'. Note that a fuzzy game is NOT a zero game, since a zero game is a win for the *second* player. The whole of Chapter 8 was built on this distinction. So we now have four types of game: positive (won by Plus irrespective of who is to start), negative (won by Minus similarly), fuzzy (won by whoever is to play first), and zero (won by whoever is to play second). A game using only neutral counters is either fuzzy or zero; a game using only personal counters is either positive, negative, or zero; a game using both personal and neutral counters may be of any type.

It is convenient to start our study of games which use both personal and neutral counters by considering the pile shown in Figure 9.10(a), in which a single positive counter is perched on top of a single neutral counter. This simple pile proves to be both interesting and important. Although it appears to favour Plus, it clearly makes a fuzzy game when played on its own; whoever is to play first can simply remove it. But if we add a single-counter neutral pile, as shown in Figure 9.10(b), we obtain a composite game which does indeed favour Plus, to the extent of being strictly positive; if Plus starts, he can remove his own counter from the left-hand pile and leave a balanced pair, whereas if Minus starts, he can only remove a complete pile, and Plus can remove the other. Yet if we add a second single-counter neutral pile, as shown in Figure 9.10(c), the composite game reverts to fuzziness, since whoever is to play first can remove the left-hand pile and leave a balanced pair.

We shall meet the game of Figure 9.10(b) again, so let us denote it by $U$. Although positive, it is very weakly positive, since the addition of a suitable neutral pile is sufficient to make it merely fuzzy. On the other hand, a composite of two such games is rather more strongly positive. Figure 9.11 shows such a composite, together with a neutral pile of arbitrary size, and it is easily seen that Plus can win irrespective of the move and irrespective of the size of the neutral pile. So if Plus's advantage in a game is comparable with $U + U$ then he wins, but if it is merely comparable with $U$ then the result may depend on the move or on the size of an accompanying neutral pile.

(a)                    (b)                        (c)

**Figure 9.10** Neutral interactions

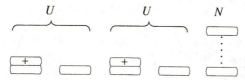

**Figure 9.11** A small but decisive advantage

The game $U$ has another interesting property: although it is positive, it is smaller than any positive number. If $x$ is any positive number, then we can obtain a number $2^{-k}$ which is less than $x$ by making $k$ sufficiently large, and we can construct a pile $P$ of measure $2^{-k}$ by placing $k$ negative counters on top of one positive. Figure 9.12(a) shows such a pile $P$, and Figure 9.12(b) shows the composite game $U+(-P)$; and it is easily seen that this composite game is a win for Minus, because Minus can refrain from removing the bottom counter of $-P$ until nothing remains of $U$. In other words, $U < P$. It does not matter how many positive counters are contained in the upper reaches of $-P$; it is the solitary negative counter at its foot that decides matters. So the measure of $U$, although positive, is smaller than any positive number, and so it must be a quantity of quite a different type. Indeed, not only is $U$ smaller than any positive number, but so is $U+U$, and so is any multiple of $U$; for the composite game $U+U+ \ldots +U+(-P)$ is still a win for Minus, because Minus can refrain from removing the bottom counter of $-P$ until nothing remains of any $U$.

We have now established everything about personalized nim that we shall need in the remainder of the chapter, but let us briefly digress and survey its remaining properties.

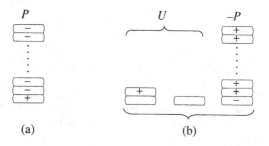

(a)                              (b)

**Figure 9.12** A very small positive game

A general nim pile can be divided into two parts: that below the lowest neutral counter, and that above and including it. We call these the 'lower' and 'upper' parts. The lower parts can be measured in the usual way, and if the sum of these measures is positive then Plus wins; he removes neutral counters as long as there are any left, and eventually comes down to a game which lacks neutral counters and has positive measure. However, if the sum of these measures is zero, the first person to remove a counter from the lower part of any pile loses, because he disturbs its measure in his opponent's favour. It follows that in this case we can ignore all the lower parts, and consider only moves within the upper parts; if a player cannot win by playing within the upper parts, he certainly cannot win by playing below them.

So let us suppose that the measures of the lower parts do indeed add up to zero. If only one pile has an upper part, the first player can now win by removing it. If two piles have upper parts, the result depends, in general, on the ownership of the lowest personal counter in each. If the same player owns the lowest counter in each, he wins; otherwise, the rules are as follows.

(a) If one upper part contains only neutral counters, the result depends on the move, the number of neutral counters, and the ownership of the lowest personal counter in the other upper part. The owner of this counter wins if he has the move, or if the number of neutral counters in the other part is at least as large as the number below his own lowest counter; otherwise his opponent wins. Figure 9.13 shows the exceptional case. Although the only personal counter in sight is positive, Minus, to move, can win by playing at the arrow; but the addition of even a single extra neutral counter on the right would give Plus an unconditional win.

**Figure 9.13** A win for the first player

(b) If each player owns the lowest personal counter in one upper part, the winner is he whose counter has the smaller number of neutral counters below it.

(c) If each player owns the lowest personal counter in one upper part and there are the same number of neutral counters below each, the winner is the first player whose opponent cannot move higher up in either pile. Only in this case do the counters above the lowest matter.

It follows that the players in a game of many piles should rush around like mad axemen and chop down each other's upper parts, priority being given to those with the smallest number of neutral counters at the bottom. Only when each of an opponent's remaining upper parts has the same number of neutral counters at the bottom need consideration be given to other than the lowest counter above.

# The nature of a numeric game

We have seen that a pile consisting entirely of positive and negative counters has a numeric measure, and that the winner of a game comprising two or more such piles can be ascertained by adding these measures according to the ordinary rules of arithmetic. On the other hand, a pile containing even one neutral counter does not. It is instructive to see why.

It is possible, following Conway, to define games by induction, using a procedure similar to that which we used when assigning Grundy values to positions in impartial games.

(a) We start by defining the game in which neither player can move at all, and we denote this game by 0.

(b) Next, we define the set $\{G_1\}$ of games in which a player can move only to 0, or not at all. There are four such games (Plus can move but Minus cannot, Minus can move but Plus cannot, both can move, or neither can move), but the last of these is 0 itself, so there are only three new ones. We have already met them, in the guise of piles containing one positive, negative, or neutral counter.

(c) We next define the set $\{G_2\}$, which is the set of games from which each player can move only to games in $\{G_1\}$, or to 0, or not at all; and this process can be continued as far as we like.

This inductive construction generates all games which can be guaranteed to terminate in a finite number of moves.

But this is not all. It is also possible, again following Conway, to define numbers themselves in a very similar way; we follow the previous definition, except that we forbid Plus and Minus to move respectively to games $P$ and $M$ such that $P \geqslant M$. Thus 0 is defined as before, as the game in which neither Plus nor Minus can move at all. The set $\{G_1\}$ now contains only two games, those in which *either* Plus *or* Minus can move to 0, and we can denote these games by $+1$ and $-1$ respectively. The apparent third element of $\{G_1\}$, in which both Plus and Minus can move to 0, is banned by our restriction, since 0 would constitute both a $P$ and an $M$ and these would trivially be such that $P \geqslant M$. The next set $\{G_2\}$ includes the game in which Plus can move only to 0 and Minus to $+1$, and we can denote this game by $+1/2$; and so on. This systematic definition yields all the finite integers and the finite binary fractions, and limiting processes then yield $1/3$, $\sqrt{2}$, $\pi$, and all the other numbers. The usual operations of arithmetic can be defined on these numbers, and they can be shown to give the answers that everyday life demands. In other words, those ordinary and rather mundane objects which we call 'numbers' are really a subset of the very much more interesting objects which we call 'games'.[2]

It follows that the condition for a game to have a numeric measure is very simple: it must not be possible for Plus and Minus to move to games $P$ and $M$ such that $P \geqslant M$. A pile consisting only of positive and negative counters satisfies this condition; whatever moves by Plus and Minus we choose, if we set up and play out the resulting composite game $P + (-M)$, we find that Minus wins, and it follows that $P < M$. On the other hand, if a pile contains a neutral counter, either player can remove this counter and everything above it, so whatever is below it constitutes both a $P$ and an $M$ which trivially satisfy $P \geqslant M$.

If Plus and Minus do have moves such that the resulting games $P$ and $M$ satisfy $P \geqslant M$ then the game cannot be represented by a single number. Some games in this class are indeed 'hard' in the sense which we shall meet in Chapter 10 (roughly speaking, it may not be possible to determine the winner without doing rather a lot of work). Nevertheless, even a game such as this may possesses measures of a sort; for example, it may be possible to quantify the advantage of making the next move. Such information is valuable when the game is played as part of a composite.

This brings us back to the heart of the matter. The ultimate object of the exercise is to identify the winner. If a game is played on its own, all that concerns us is whether it is positive, negative, fuzzy, or zero; the more precise measure which we have sought in this chapter

---

[2] The reader who seeks a full treatment of this fascinating subject should consult either *On numbers and games* or *Winning ways for your mathematical plays*.

is important only when the game forms part of a composite. Some games, such as nim, start out as composites, and others, such as kayles, usually develop into composites in the normal course of play. But there are many games, chess and draughts (checkers) being obvious examples, which usually remain as single coherent entities throughout play. It is true that there are endgame situations in chess and draughts which have the character of composites, but these occur only infrequently and are in no sense typical of the game as a whole. The theory of numerical measure has little relevance to such games.

# The measure of a feeble threat

We have seen that a numeric game can have as small a measure as we like; and we have seen that there are games, such as $U$ in Figure 9.10(b), which are positive but smaller than all positive numeric games. Yet there are games which are smaller still.

A habit sometimes adopted by children who are faced with inevitable loss at chess is to prolong the game by 'spite checks': sacrificial attacks on the winner's king, which merely waste time and have no effect on the ultimate result of the game. (Children usually grow out of this habit; computers, in my experience, do not.) Such a check is a special case of a feeble threat, feeble because it can easily be countered, and equivalents occur in many games. Yet for all their apparently trivial nature, they have some interesting mathematical properties.

The nature of feeble threats can most easily be examined by inventing a 'pileworm', which may be regarded as the larval stage of a pile. If Plus possesses a pileworm, he is allowed only to convert it into a pupa; then, on a subsequent move, he may convert this pupa into a pile of his own counters (one for each segment of the original worm, not counting the head), which he can then remove in the normal way. Minus can remove the embryonic pile while it is still a worm or while it is a pupa, but he cannot touch it once it has become a pile. A pileworm therefore constitutes a very feeble threat indeed. As long as it remains a worm, there is no immediate danger; and to convert it into a pupa takes a move, which the opponent can use to remove it.

Yet for all that, it has an existence. On its own, a Plus pileworm is a *negative* game.[3] Minus, to play, can remove it as a worm; Plus, to

---

[3] Readers who find difficulty in comprehending the negative nature of a Plus pileworm may care to consider the analogy of a child in everyday life. While it is very young, a child is a *liability* to its family, since it consumes resources without giving anything in return. It only becomes an asset as it grows older.

play, can merely pupate it, and Minus can then remove it as a pupa. But it is not a very large game. It is very much smaller than our game $U$ in Figure 9.10(b). We saw earlier that $U$ was positive; Figure 9.14 shows a composite game consisting of $U$ and a number of Plus pileworms, and it is easy to show that this composite is still positive. Plus, to play, pupates a worm, and Minus must get rid of it before it hatches. Plus then pupates another, and so on. Minus, to play, would merely accelerate matters by removing a worm, so he plays in $U$; but Plus can still pupate all his worms, and force Minus to waste time removing the pupae, before making his winning reply in $U$. The point is that a pupa, *if allowed to hatch*, would dominate everything else, so hatching cannot be allowed. It does not matter how many segments the worm contains; even a 'fractional' worm that would ultimately yield a pile of measure $2^{-n}$ cannot be allowed to hatch.

**Figure 9.14** Even smaller games

We can say more. Let $X_n$ be a positive game which must terminate in at most $n$ moves, and let $P_{n-1}$ be the Plus pileworm containing $n-1$ segments; then the game $X_n + P_{n-1}$ is still positive, because Minus cannot allow the worm to hatch. (The resulting pile would give Plus $n-1$ moves, and Minus would have at most $n-1$ moves left in $X_n$ after playing his first.) In other words, a pileworm containing $n-1$ segments is smaller than *any* positive game which is guaranteed to terminate within $n$ moves.

One final frivolity. Suppose that we have a game consisting of pileworms alone. The worm with the *fewest* segments now dominates. Figure 9.15 shows one single-segment Minus pileworm (which is a positive game) faced with several double-segment Plus pileworms. Plus, to play, pupates a worm as usual. Minus, to play, pupates his worm in the hope that Plus will get rid of it, since everything else in sight is negative; but Plus pupates one of his own in reply, and Minus must remove it since it will dominate his own if it is allowed to hatch. So Plus gets rid of all his own worms before he has to take Minus's pupa. The worm with the fewest segments is actually the largest worm.

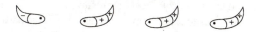

**Figure 9.15** The war of the worms

# Infinite games

So far, we have been careful to confine our attention to finite games. To round off the chapter, let us look at a few infinite games.

The simplest infinite games are infinite piles. We give three examples. The most straightforward of all is the infinite pile of positive counters, shown in Figure 9.16(a). Plainly, the possession of such a pile enables Plus to win against any combination of finite piles. But if Minus has a similar pile? If there is no other pile in play, the second player wins, since his opponent's first move must reduce his pile to finite proportions and he can choose his own first move so as to leave a larger pile. If we call the infinite pile of positive counters $Z$, this amounts to the observation that $Z + (-Z) = 0$.

Almost as straightforward is the pile shown in Figure 9.16(b). This is clearly positive, but it is a pile of very small measure; indeed, it is smaller than any pile of finite measure $2^{-n}$. Since it lies at the other end of the scale of numbers from $Z$, we might perhaps call it $A$. Yet $A$, infinitesimal though it might seem, is still larger than any multiple of $U$. The argument given earlier in the chapter, wherein we showed that the composite game $U + U + \ldots + U + (-P)$ was a win for Minus, remains valid if we replace $-P$ by $-A$; as before, Minus can refrain from removing the bottom counter of $-A$ until nothing remains of any $U$.

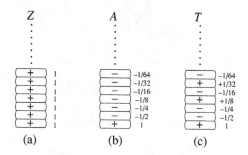

**Figure 9.16** Some infinite piles

The third and perhaps the most interesting of our examples is the pile $T$ shown in Figure 9.16(c), which is the pile obtained by applying our previous rule to the number 1/3. Now when we wished to test the proposition that the game $G$ in Figure 9.5 had measure 1/2, we set up the composite game $G+G+(-P_1)$ and showed that it was a win for the second player. This suggests that if $T$ truly has measure 1/3 then the game $T+T+T+(-P_1)$ should be a win for the second player. Well, is it? Figure 9.17 shows typical play with Plus starting. The first three moves are essentially arbitrary; Plus must make his first move somewhere in one of the piles $T$, Minus should reply higher up in a second pile $T$, and Plus should now play still higher up in a third. The rest is automatic, and we see that Minus can indeed win if Plus starts. A similar analysis shows that Plus can force a win if Minus starts, so we conclude that the pile $T$ does indeed have measure 1/3. Readers with a taste for such things may also care to verify that the pile in which the counters alternate $+-+-+-\ldots$ right from the bottom has measure 2/3.

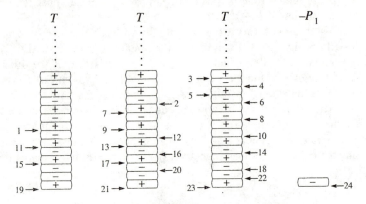

**Figure 9.17** A pile worth a third of a counter

But it can be argued that infinite piles are not truly infinite, merely unbounded. It may be impossible to say in advance how long a game will last, but ultimate termination is certain. The game of Figure 9.17 is triply unbounded, but ultimate termination is certain even here; once we have broached a pile, we are certain to finish it, and we must then broach another. To obtain a truly infinite game, we must allow a player to return to a previous state. The simplest game in this class is the single-state game shown in Figure 9.18(a), in which Plus is

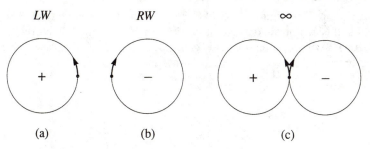

**Figure 9.18** Loopy games

allowed to return but Minus is not allowed to move. Since the conventional positive direction of angular movement is anticlockwise, we might call this game *LW* for 'left wheel'. Now the game *LW* presents theoretical difficulties, since it cannot be defined by induction; an inductive procedure defines a game by allowing the players to move to games that have already been defined, and Plus's only move in *LW* is to *LW* itself. Yet *LW* clearly exists, and its properties are straightforward albeit somewhat drastic. If *F* is any finite game, $LW + F$ is an easy win for Plus, since he can keep moving in *LW* and wait for Minus to run out of moves in *F*. Indeed, even $LW + (-Z)$ is a win for Plus, because Minus's first move reduces $-Z$ to a finite pile. More generally, let *X* be any game, terminating or not; then Plus certainly doesn't lose in $LW + X$, since he can keep moving in *LW*. In other words, nothing can beat *LW*; in a sense, it is the largest game that exists.

Apparently complementary to *LW* is the game *RW* ('right wheel') in which Minus can pass, as shown in Figure 9.18(b). But there is a trap. When we introduced complements earlier in the chapter, we were careful to say that the relation $X + (-X) = 0$ applied only if the rules of the game guaranteed termination. No such guarantee applies here; quite the reverse, in fact. Indeed, we *cannot* say that $LW + RW = 0$, because this would mean that $LW + RW$ is a second-player win, and we have already seen that Plus doesn't lose in $LW + X$ whatever *X* might be. So we see that *Z* and *LW* represent quite different kinds of 'infinity'. Not only is *LW* bigger than *Z*, in the sense that $LW + (-Z)$ is a win for Plus, but we do have $Z + (-Z) = 0$ whereas we don't have any game '$-LW$' such that $LW + (-LW) = 0$.

But if $LW + RW$ doesn't equal 0, what does it equal instead? The answer is the game shown in Figure 9.18(c), in which both players can loop. This figure looks like the conventional infinity symbol $\infty$, and this is not an inappropriate name for the game. Whatever the game $X$, the combination $\infty + X$ cannot be lost by either player; even if he cannot move in $X$, he can always move in $\infty$.

# IO

# WHEN THE COUNTING HAS TO STOP

To complete our study of games of pure skill, we look at what happens when mathematical analysis fails to provide a satisfactory answer; and we consider the fundamental paradox, that a game of pure skill can be played in competition only as long as the players are sufficiently ignorant.

## The symptoms of a hard game

We start by looking at the meaning of the term 'hard' as applied to mathematical games, and at the practical implications of hardness.

In everyday usage, the word 'hard' is an imprecise term of abuse, and means only that the speaker finds it difficult or impossible to perform what is demanded. In the theoretical analysis of a finite game, the question of impossibility does not arise, while 'difficult' is a comment on the player as much as on the game. We need a definition which relates solely to the game itself but is consistent with the connotations of everyday life.

One way of achieving such a definition is to generalize the game in some way, and then to observe how the amount of computation needed to analyse a specific 'instance' of it (typically, the play from a given starting position) increases with the amount of numerical information needed to define this instance. If the increase is merely linear, the game is relatively easy; if it follows some other polynomial of low degree, the game may not be too bad; if it is exponential, the game is likely to be difficult; and if it is worse than exponential, the game is hard indeed.[1]

But we cannot play a generalized game; we can only play a specific instance. It does not help us to know that the amount of computation

---

[1] For a fuller discussion, see *Winning ways for your mathematical plays* 218-20; for a fuller discussion still, see the bibliography cited therein.

demanded by a generalized game increases only linearly, if the specific instance which we are required to play is too large and complex for us to analyse; while, conversely, many instances of theoretically hard games are in practice quite tractable. Nevertheless, if a generalized game is hard then some of its instances must reflect its hardness, and this reflection typically takes the form of a lack of apparent pattern in the play. We cannot use logic to diagnose short cuts, but must examine variations in depth.

An instructive example is given by chess. That a game as complex as chess should be hard is not surprising, but in fact its hardness may become perceptible even in apparently straightforward endgame situations. Suppose that each side has a king and a queen, and that White also has one or more pawns. In general, White will now attempt to win by promoting a pawn to a second queen, while Black will seek to draw by subjecting the White king to 'perpetual check' (repeated attacks from which he can never escape). These objectives are exemplified by Figures 10.1 and 10.2.

In Figure 10.1, White has a simple win. Although he is in check, he can promote his leading pawn, which leaves Black with no sensible further check; and the rest is easy. Furthermore, the White pawns form a shelter towards which White could have aimed earlier in the play, so the play leading up to Figure 10.1 may also have been straightforward.

In Figure 10.2, on the other hand, Black has a draw. Even after White has promoted his pawn to a second queen, his king is open to attack from three directions (on the rank, on the file, or on the diagonal) and he can only cover two of them at once. So if White leaves the file uncovered, Black checks on $A$; if White leaves the diagonal uncovered, Black checks on $B$; and if White leaves the rank uncovered, Black checks on his present square. Black can keep this up indefinitely, and White can never escape.

But there is an intermediate class of position in which White can prevent an immediate repetition but apparently cannot prevent Black from continuing to check somewhere or other. Such a case is illustrated in Figure 10.3. This particular position can in fact be won by White, as was first shown in a computer analysis by E. A. Komissarchik and A. L. Futer, but there is no apparent pattern to the winning manoeuvres, and they could not have been discovered without exhaustive analysis. Figure 10.4 shows the moves required of the White king in reply to Black's best play, and the contorted sequences are characteristic of the play that is required in instances of hard

**Figure 10.1** Chess: an easy win

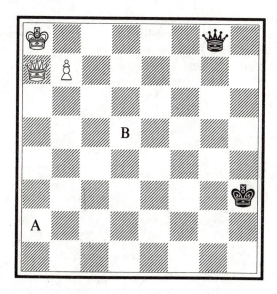

**Figure 10.2** Chess: an easy draw

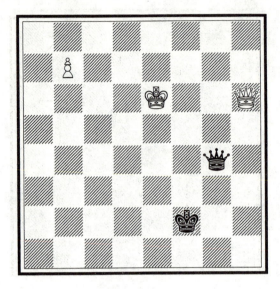

**Figure 10.3**  Chess: a very difficult win

games. Subsequent computer analyses by K. L. Thompson and A. J. Roycroft have evaluated all chess positions featuring king, queen, and one pawn against king and queen, and have discovered some which require winning sequences even longer and more convoluted than those demanded by Figure 10.3.[2]

<hr />

[2] The Thompson-Roycroft results are recorded in *Roycroft's 5-man chess endgame series* (Chess Endgame Consultants and Publishers, London, 1986). This publisher also produces the chess endgame study magazine *EG* which records advances in this field. The Komissarchik-Futer solution to Figure 10.3 was originally published in the paper 'Ob analize ferzevovo endshpielya pri pomoschi EVM', *Problemy kibernetiki* **29** 211–20 (Nauka, Moscow, 1974); the main line is reprinted with additional commentary in *EG* **4** 239–41 (November 1979), and more briefly in *The Oxford companion to chess* (D. V. Hooper and K. Whyld, 1984, entry 'Basic endgame'). The Thompson-Roycroft solution to Figure 10.3 is in fact slightly different from the Komissarchik-Futer solution, and suggests that the latter is not quite optimal for either player; the discrepancies are not such as to cast serious doubt on the overall conclusions, but they do make the point that a mathematical proof depending on exhaustive analysis by computer should really be regarded as provisional until it has been confirmed by an independent worker.

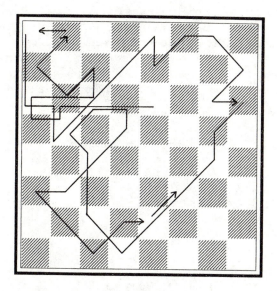

**Figure 10.4** The odyssey of the White king (Kommisarchik and Futer). If Black plays his best moves, this is what White must do in order to win. There is no question of systematic progress towards a perceptible goal, and only exhaustive analysis by computer could have discovered the solution

# When you know who, but not how

If we are playing an instance of a hard game, we may not be able to find the best play within a reasonable time, and the same may be true even of instances of games which are not hard. However, some games have the remarkable property that we can diagnose the winner without being able to say how he wins. Most of these games are now well known, but they bear repetition.

## (a) The completion of patterns

There is a class of games, typified by the humble noughts and crosses (tic-tac-toe), in which each player occupies a cell on a board in turn and seeks to complete a specified pattern before his opponent does so.

Hex, invented by Piet Hein, is a game in this class. It is played on a board such as that shown in Figure 10.5, the actual size of the board being a matter for agreement between the players. Each player takes a pair of opposite sides, and his move is to place a man in any unoccupied cell, his objective being to form a continuous chain between his two sides. Thus Figure 10.5 shows a win for Black.

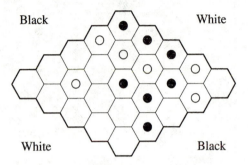

**Figure 10.5** Hex: a win for Black

It is not difficult to see that a game of hex can never be drawn. After the last cell has been filled, we can start at one White edge, marking all White cells now joined to that edge. If the marked cells extend right across the board, they form a winning chain for White; if they do not, they must be blocked by a solid line of Black cells, and these form a winning chain for Black.

It is also not too difficult to see that the first player has a certain win, and the proof, due to J. L. Nash, appears to have been the first application of the argument now known as 'strategy stealing'. Suppose that the second player has a winning strategy. Now let the first player make his first move at random and then follow this strategy himself, substituting another random move whenever the strategy requires him to play in a cell already occupied. The extra man placed at random does the first player no harm, so this procedure must guarantee him a win. Hence no winning strategy for the second player can exist.

A similar argument applies to every game of this type in which a draw is impossible and an extra man is never harmful. Bridg-it, invented by David Gale, is a case in point. It is played on a pair of interlaced lattices such as are shown in Figure 10.6, and a player's move is to join two adjacent points of his own colour. His objective, as in hex, is to connect his own two sides of the board. The game

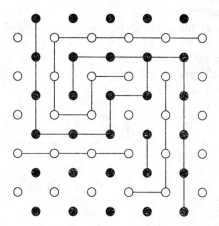

**Figure 10.6** Bridg-it: another win for Black

cannot be drawn and an extra line never does any harm, so the first player always has a win.[3]

Indeed, a weaker form of the argument applies even to noughts and crosses itself. This game can be drawn, so strategy stealing does not prove that the first player can win; but it does prove that he can prevent the second player from doing so.

## (b) Chomp

This is a game of quite a different character. It is played on a rectangular array which is initially full of men, and the move is to take a rectangular bite from the top right-hand corner (Figure 10.7). The objective is to force the opponent to take the last man. The game is due to Gale, the name (I think) to BCG.

If the rectangle comprises a single man only, the first player loses trivially; if it contains more, he can always win. The proof is due to Gale himself. If the first player can win by removing just the top right-hand man then there is nothing to prove. Alternatively, if the first player cannot win by removing just this man then the second player must have a winning reply, by removing everything upwards and rightwards from some other man; but this move would have

[3] There is a difference, however. Bridg-it possesses a known and reasonably straightforward winning strategy for the first player, whereas generalized hex is hard. See *Winning ways for your mathematical plays* 680–2.

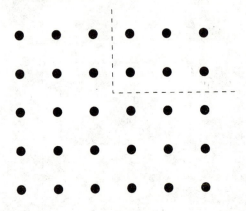

**Figure 10.7** Chomp

removed the original top right-hand man had it still been present, so the first player could have produced the same position by making this move immediately.

## (c) *Sylver coinage*

More subtle than any of the previous games is Conway's game of sylver coinage.[4] In this game, each player in turn names a positive integer, which is to be minted as a unit of coinage; but he may not name a unit which can already be achieved as a sum of existing units. For example, if '5' has already been named, a player may not name any multiple of 5; if '5' and '3' have both been named, a player may not name any number higher than 7. The first player to name '1' loses.

This game also yields to a strategy-stealing argument, due to R. L. Hutchings.

(i) Suppose that we have a position in which only a finite number of integers remain to be named, and that the highest of these will become unavailable if any lower integer is named. There is now a winning move for the first player, by a similar argument to that used

---

[4] It is a slightly moot point whether the name should be 'sylver' or 'Sylver' coinage. In accordance with the modern style, I have omitted the initial capitals which used to be customary when naming games, but sylver coinage is punningly named in honour of J. J. Sylvester, whose theorem plays an important role in its theory. When does a proper name cease to be proper? The analogy with scientific units suggests that the capital should be dropped, but the matter is not entirely clear.

above; if he cannot win by naming the highest remaining integer, he can win by making the second player's winning reply.

(ii) If *a* and *b* have no common factor, the naming of *a* and *b* makes unavailable all except a finite number of integers, the highest being $ab - a - b$; furthermore, unless this highest integer is 1 (which happens only if *a* and *b* are 2 and 3), it will be made unavailable by the naming of any lower integer. This is Sylvester's theorem of 1884 (*Math. Quest. Educ. Times* **41** 21).

(iii) Therefore the first player can win by naming any prime greater than 3.

This beautiful proof seems to me to be in quite a different class from others involving strategy stealing. It even specifies an infinite number of winning moves for the first player. Yet in spite of this, it is not very helpful. You, being knowledgeable, start by naming 5; I, following the sound policy of complicating a lost position in the hope that you will make a mistake, reply by naming 999 999 999; and your computer has to do a lot of work before you can make your next move with confidence.

# The paradox underlying games of pure skill

Inability to determine the result of a game without impracticably detailed calculation does have one apparent compensation: it leaves the game alive for competitive play. Once a game of pure skill has been fully analysed, it is competitively dead. It is only the ignorance of players that keeps games such as chess alive at championship level.

Yet it must be questioned whether ignorance really provides a sound basis for the perpetuation of a game of skill. Draughts (checkers) is rarely played at the highest level, because good players find it too easy to hold a game to a draw. Chess has not yet reached this stage, but the recent Karpov–Kasparov matches for the world championship have not been encouraging. Their first match, in 1984, should have gone to the first player to win six games, but it was controversially abandoned by the authorities when the score was 5–3 with 40 draws. Subsequent matches have been restricted to 24 games, the defending champion retaining his title in the event of a tie, and such a formula should reduce the number of draws since there is always one player who must take risks in order to gain ground; yet even so, nearly two thirds of the games played in the 1985-7 matches were drawn. Overall, Kasparov currently leads by 17-16 with 87 draws, and this is not a recipe for the maintenance of popular enthusiasm.

Why, then, do people retain an interest in such a game, whether as performers or as spectators? Perhaps there are three reasons. Some enthusiasts are merely competitive; some seek beauty; and some seek truth.

Those who are merely competitive are of no interest to us. Sadly, however, they are all too prevalent; at every level of the game, there appear to be players whose sole concern is to prove themselves better than their neighbours. Persons of this class do not care that their 'triumphs' merely reflect the limitations of their opponents. This excessive competitiveness is frequently fostered by the ill-considered ambition of parents and schoolteachers; and also by national pride, since success at a game is all too often represented as demonstrating the superiority of a particular nation or political ideology. Politically motivated persons should be wary of quoting these words in argument, since propagandists for many nations and ideologies have been guilty of this folly.

Those who seek beauty are quite different. It is almost intrinsic that a game of skill may contain moments of beauty; a blunder by the opponent may be necessary to set the scene, but the exploitation of the situation once it has arisen is another matter entirely. So the games containing beautiful incidents find their way into anthologies, where they can be studied and replayed for pleasure by ordinary enthusiasts; and it is not a very large step from the exploitation of casual accidents to the search for beautiful play as an end in itself, regardless of whether a plausible sequence of blunders might produce the necessary position in an actual game. Even the highest competitive levels contain players who have taken this step. Réti and Mattison each had a victory over a reigning world champion to his credit, yet neither was obsessed by competition for its own sake; each could turn aside from the hurly-burly of tournament play, and spend time discovering play such as is shown in Figures 10.8 and 10.9. The deeper complexities of chess have been avoided in this book, but every reader who plays the game at all will enjoy getting out a set and playing these through.

Those who analyse games mathematically come into the third category: seekers after truth. The truth may indeed turn out to be beautiful, as in most of the games which we have considered here, but we do not know this before we start; it is sought as an end in itself. This is the motivation that drove Bouton to discover the mathematical theory underlying nim, and Thompson and Roycroft to seek a resolution of all chess endings with queen and pawn against queen. In the words of Hillary's justification for climbing Everest, we seek the truth because it is there; and the implications for competitive play are not of interest.

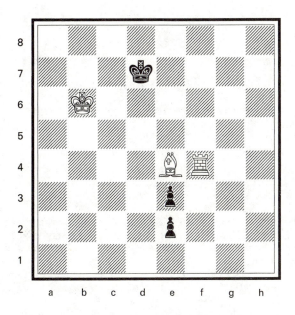

**Figure 10.8** Chess: an endgame study by Réti, as modified by Rinck. White will win if he can capture the pawns without losing his rook, and after 1 Bf5+ Kd6(d8) 2 Rd4+ Ke7 3 Re4+ he appears to have done so. But Black has a subtle resource: 3 . . . Kd8. Now 4 RxP Pe1 = Q 5 RxQ will give stalemate, and surely White has no other way of making progress? But indeed he has, in the remarkable move 4 Bd7. Like all really brilliant moves, this looks at first like a beginner's blunder, since it gives away the bishop and seems to do nothing useful; but if 4 . . . KxB then 5 RxP Pe1 = Q 6 RxQ and there is no longer a stalemate, and if 4 . . . Pe1 = Q instead then the elegant continuation 5 Bb5 shields the White king, and Black can avoid immediate mate only by sacrificing his queen

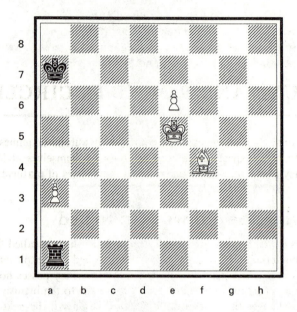

**Figure 10.9** Chess: an endgame study by Mattison. The obvious 1 Pe7 does not win, since Black can sacrifice rook for pawn by 1 ... Re1+ 2 Kf6 RxP 3 KxR; White may be a bishop and a pawn ahead, but he has no way of driving the Black king out of the corner. So White must try 1 Be3+ Kb7 2 Pe7, and 2 ... RxP now seems unanswerable; if 3 Pe8=Q then Black plays 3 ... RxB+ and 4 ... RxQ, and if the bishop moves to safety then Black can draw by 3 ... Ra8 and 4 ... Re8. There is just one way to win: 3 Ba7. This lovely move prevents 3 ... Ra8, and neither rook nor king can safely capture the interloper; if 3 ... RxB then 4 Pe8=Q and queen wins against rook, while if 3 ... KxB then 4 Kf4 (or Kd4) and White will promote as soon as Black's checks have run out. But Black can fight on by 3 ... Ra1, aiming for e1. Now 4 Pe8=Q Re1+ fails to win, as does 4 Ke4 KxB 5 Pe8=Q Re1+; White must play the subtle 4 Kf4. The intended 4 ... Re1 can now be met by 5 Be3, and if Black tries to dislodge the king by 4 ... Rf1+ then White sacrifices the bishop again, and this second sacrifice allows the king to cramp the rook: 5 Bf2 RxB+ 6 Ke3 Rf1 7 Ke2 (quickest) and the pawn will safely be promoted

# I I

# ROUND AND ROUND IN CIRCLES

In this final chapter, we examine some purely automatic games. They prove to be unexpectedly interesting, both in themselves and in the light they throw on certain fundamental paradoxes of mathematics.

## Driving the old woman to bed

There is a well known card game for children which is called 'beggar your neighbour' or 'drive the old woman to bed'. Each player has a stack of cards, and plays in rotation to a trick. Play passes normally as long as the cards played are plain cards (two to ten inclusive), but the rule changes the moment a court card is played; the next player must now play repeatedly until *either* he has laid down a certain number of plain cards (one if the court card was a jack, two if a queen, three if a king, four if an ace) *or* he plays a court card himself. In the former case, the player who played the court card picks up the trick, puts it on the bottom of his stack, and leads to the next trick; in the latter, the next player in turn must play to the new court card in the same way. A player whose stack is exhausted drops out, and the winner is the last to be left in play.

This is an automatic game with no opportunity for skill, which is why it is so suitable for family play. From a mathematical point of view, however, it raises some interesting questions.

(a) How long is a game likely to last?

(b) Can a game get into an infinite loop?

(c) If it cannot, can we hope to prove this?

I know of no complete answer to the first of these questions, but Table 11.1, which summarizes a computer simulation of slightly over ten thousand two-player deals, may throw some light. The interpretation of this table is straightforward; for example, the first row states that no game terminated on the first or second trick, 2

**Table 11.1** Beggar your neighbour: a simulation by computer

| Tricks | Occurrences | | | | | | | | | | R(%) | $D_{20}$(%) |
|---|---|---|---|---|---|---|---|---|---|---|---|---|
| 1–10 | 0 | 0 | 2 | 21 | 84 | 190 | 250 | 237 | 243 | 265 | 87.5 | 49.4 |
| 11–20 | 301 | 274 | 273 | 273 | 258 | 263 | 228 | 234 | 236 | 237 | 62.5 | 49.5 |
| 21–30 | 237 | 219 | 198 | 209 | 192 | 161 | 163 | 176 | 178 | 143 | 44.3 | 49.4 |
| 31–40 | 140 | 167 | 144 | 137 | 144 | 144 | 115 | 112 | 103 | 110 | 31.5 | 50.9 |
| 41–50 | 81 | 112 | 95 | 102 | 110 | 87 | 74 | 98 | 92 | 90 | 22.4 | 50.7 |
| 51–60 | 81 | 75 | 78 | 66 | 92 | 70 | 58 | 63 | 74 | 55 | 15.5 | 51.0 |
| 61–70 | 65 | 48 | 52 | 39 | 46 | 56 | 36 | 40 | 41 | 36 | 11.0 | 50.9 |
| 71–80 | 40 | 36 | 28 | 45 | 47 | 37 | 35 | 29 | 23 | 35 | 7.6 | 50.4 |
| 81–90 | 30 | 18 | 22 | 27 | 25 | 20 | 27 | 15 | 20 | 20 | 5.4 | 54.0 |
| 91–100 | 24 | 23 | 13 | 16 | 18 | 17 | 19 | 14 | 17 | 10 | 3.8 | 52.6 |
| 101–110 | 18 | 9 | 15 | 16 | 11 | 12 | 11 | 14 | 14 | 11 | 2.5 | 50.2 |
| 111–120 | 6 | 7 | 12 | 7 | 5 | 7 | 11 | 5 | 7 | 6 | 1.8 | 50.5 |
| 121–130 | 8 | 7 | 8 | 6 | 4 | 3 | 4 | 3 | 8 | 5 | 1.2 | 50.8 |
| 131–140 | 3 | 3 | 4 | 6 | 6 | 4 | 3 | 4 | 3 | 1 | 0.9 | 51.6 |
| 141–150 | 4 | 4 | 2 | 6 | 2 | 2 | 3 | 4 | 0 | 1 | 0.6 | 58.7 |
| 151–160 | 2 | 4 | 1 | 1 | 0 | 2 | 3 | 1 | 4 | 1 | 0.4 | 61.4 |
| 161–170 | 3 | 3 | 2 | 1 | 3 | 2 | 0 | 0 | 2 | 2 | 0.3 | 50.0 |
| 171–180 | 0 | 0 | 1 | 1 | 2 | 0 | 0 | 1 | 3 | 1 | 0.2 | 47.1 |
| 181–190 | 1 | 0 | 1 | 0 | 0 | 0 | 2 | 0 | 0 | 0 | 0.1 | 46.2 |
| 191–200 | 1 | 1 | 1 | 0 | 0 | 0 | 0 | 1 | 0 | 0 | 0.1 | — |
| 201–210 | 0 | 0 | 0 | 0 | 2 | 0 | 0 | 0 | 0 | 0 | 0.1 | — |
| 211–220 | 0 | 1 | 0 | 0 | 0 | 0 | 0 | 1 | 0 | 0 | 0.0 | — |
| 221–230 | 0 | 0 | 0 | 0 | 0 | 0 | 1 | 1 | 0 | 1 | 0.0 | — |
| 231–240 | 1 | 0 | 0 | 0 | 0 | 0 | 0 | 0 | 1 | — | — | — |

10 311 pseudo-random deals were simulated by computer. Column $R$ gives the percentage of games that had not terminated after the stated number of tricks; column $D_{20}$ gives the percentage of these games that did terminate within a further 20 tricks. The longest deal took 239 tricks.

games terminated on the third, 21 on the fourth, and so on; that 87.5 per cent of the games had not terminated after ten tricks; but that 49.4 per cent of these had terminated within a further twenty tricks. The approximate constancy of this last column is the most interesting feature of the table, because it suggests that the game has a 'half-life' of about twenty tricks when played with a standard 52-card pack (in other words, if a game is down to two players but has not yet finished, the probability is about evens that it will still not have finished within a further twenty tricks). This implies that the beggaring of neighbours is by no means an ideal pursuit for small children immediately before bedtime, but no doubt resourceful parents have ways of coping.

It is even less clear whether the game can loop. The simulation recorded in Table 11.1 failed to produce a loop, which suggests that one is very unlikely in practice (and even if an appropriate starting arrangement were to be dealt, a misplay would probably destroy the loop sooner or later). On the other hand, loops can certainly occur with reduced packs. For example, suppose that the pack is reduced to six cards, four plain and two jacks, that there are two players, and that the cards are initially distributed as in Figure 11.1. If *A* starts, he plays a plain card, *B* plays a jack, *A* plays another plain card, and *B* takes the trick. *B* now has four cards to his opponent's two, but he has to lead to the next trick, so the original situation has been exactly reversed. A second trick restores the original situation, and so on.

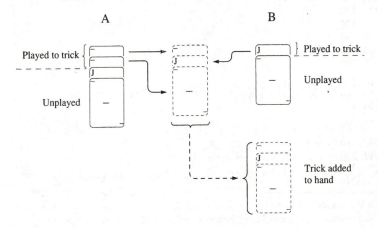

**Figure 11.1** Driving the old woman round in circles

A loop as simple as this cannot occur with a standard pack, because the proportion of plain to court cards is different; there are only 36 plain cards, whereas the four jacks, queens, kings, and aces create a total demand of 40. So from time to time, one court card must be played on another, and the trick becomes more complicated. But this does not of itself preclude a loop. In the situation shown in Figure 11.2, *A* leads a plain card to the first trick, *B* plays a plain card, *A* plays his first jack, and *B* plays another plain card, so *A* wins the first trick. *A* then leads his other jack to the second trick, but *B* plays his own jack, and *A* can only play a plain card; so *B* wins the second

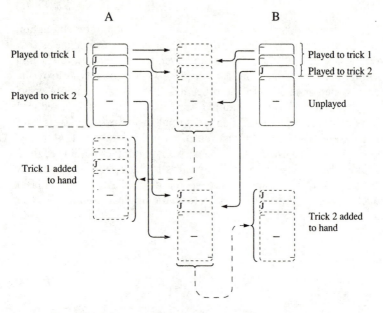

**Figure 11.2** A more complicated circuit for the old woman

trick, capturing a jack from *A* in the process. This again exactly reverses the original situation.

Whether a loop does exist for a full 52-card pack is not known to me. If it does, the problem can be resolved by finding one; but random dealing is unlikely to be profitable, and even systematic exploration is not certain to succeed within a reasonable time.[1] Yet if no loop exists, the proof is likely to be even more difficult. Since the pack is finite, we can in principle try every case, but such a task is well beyond the scope of present-day computers, and we shall see in the next two sections that there are games where even this miserable option is denied to us. Other options appear few indeed. The most common

---

[1] A crude probabilistic argument is revealing. Suppose that a pack can be arranged in a large number $N$ of different ways, that a proportion $p$ of these arrangements are defined as 'terminal', that each of the others has a defined successor, and that no two arrangements have the same predecessor. If the successors have been determined at random, the probability that a randomly chosen starting arrangement leads into a loop can be shown to be approximately $(1-p)/pN$. A half-life of around twenty corresponds to $p \approx 0.035$, upon which this probability reduces to approximately $28/N$. On the other hand, the probability that there is a loop somewhere can be shown to exceed 0.9. So we have a needle in a haystack; a loop is very likely to exist, but almost impossible to find.

way of proving that a process cannot repeat indefinitely is to show that some property is irreversibly changed by it; for example, the non-existence of perpetual motion machines is a consequence of an experimental law of thermodynamics known as 'the law of increase of entropy'. But it is hard to envisage a property that is irreversibly changed by a run through a standard 52-card pack but not by a run through the restricted packs of Figures 11.1 and 11.2.

# Turing games

The essential ingredients of an automatic game are a set of movable objects and an automaton which can assume various states. In each state, the automaton examines the set of objects, moves them, and assumes a new state. For example, the automaton in 'drive the old woman to bed' can assume four states: $A1$ ($A$ to play, no court card having been played to the current trick), $A2$ ($A$ to play, at least one court card having been played), and $B1$ and $B2$ ($B$ to play similarly). Its action in state $A1$ is to stop if $A$'s stack is empty, and otherwise to move $A$'s top card to the trick and to assume state $B1$ or $B2$ according as this is a plain or court card. Its action in state $A2$ is more complicated, since it must compare the number of plain cards on top of the trick with the number demanded by the most recent court card, but there is no difference in principle.

Another example is provided by a Turing game. Such a game features a line of coins, together with a robot which runs up and down turning them over. The robot is extremely simple; in each state, it can only examine the current coin, turn it over if required, and move one step to the left or right. Thus the action of a robot in state 1 might be specified as follows: if the current coin is a head, leave it alone, move one step right, and remain in state 1; if a tail, turn it over, move one step left, and assume state 2. The outcome of the game is completely determined by the actions of the robot in each state and by the initial orientations of the coins.[2]

The progress of a Turing game is most easily illustrated by an example. Suppose that the robot is indeed in state 1 as defined above, and that it is in the middle of a row of heads. It therefore sees a head,

[2] Some readers may recognize a 'Turing game' as a thinly disguised form of the 'Turing machine' which was invented by Alan Turing (1912–54) to resolve a fundamental problem in the theory of computation. Perhaps it is selling Turing short to present his conception in the guise of a game, but the logic is unaffected, and the present form is undoubtedly the more suitable here.

so it leaves it alone, steps right, and remains in state 1. This continues until the robot moves off the end of the row, when it sees a tail and turns it over. So the robot adds a head to the right-hand end of the current row, and then assumes state 2.

Table 11.2 shows a complete program of actions. Suppose that the line of coins contains a single row of heads, and that the robot is somewhere within this row (Figure 11.3). In state 1, as has been shown above, the robot adds a head to the right-hand end of the row. We shall see in a moment that this new head provides a seed from which a new row is generated. Being now in state 2, the robot turns over the right-hand member of the previous row, steps to the next coin on the left, and assumes state 3. The coin at which the robot is now looking is another head, so it steps back to the right and assumes state 4. In this state, it runs over the new tail and then the new head; in state 5, it adds a further new head on the right; in state 6, it runs back over the new row; and it then reverts to state 2. A pattern now begins to emerge, in that the robot is systematically removing a head from the right-hand end of the old row and adding one to the right-hand end of the new. This process continues until there is nothing left in the old row, upon which the robot finds a tail while in state 3. This tail causes it to assume state 7, in which it turns back everything

**Table 11.2** A specification of a simple Turing game

| | Seeing a head | | | Seeing a tail | | |
|---|---|---|---|---|---|---|
| State | Action | Step | New state | Action | Step | New state |
| 1 | Leave | Right | 1 | Turn | Left | 2 |
| 2 | Turn | Left | 3 | Leave | Left | 2 |
| 3 | Leave | Right | 4 | Leave | Right | 7 |
| 4 | Leave | Right | 5 | Leave | Right | 4 |
| 5 | Leave | Right | 5 | Turn | Left | 6 |
| 6 | Leave | Left | 6 | Leave | Left | 2 |
| 7 | Turn | Right | 8 | Turn | Right | 7 |
| 8 | Leave | Right | 8 | Turn | None | – |

**Figure 11.3** The start of a Turing game

**Figure 11.4** The finish of the game

in the old row and also the original seed coin for the new; and in state 8, it then adds a final coin to the end of the new row. When the robot eventually rests its weary limb, we find that it has exactly duplicated the original row (Figure 11.4).

Table 11.3 shows a more complicated program, which assumes that the line initially contains two rows of heads, separated by a single tail, and that the robot starts within the right-hand row (Figure 11.5). We omit the details; suffice it to say that in states 1-6, the robot duplicates the row in which it starts; in states 7-14, it adds a copy of the second row; and in states 15-18, it tidies up. The overall effect is to form a new row whose length is the sum of the lengths of the original rows (Figure 11.6).

**Table 11.3** A more complex Turing game

| | Seeing a head | | | Seeing a tail | | |
|---|---|---|---|---|---|---|
| State | Action | Step | New state | Action | Step | New state |
| 1 | Leave | Right | 1 | Turn | Left | 2 |
| 2 | Turn | Left | 3 | Leave | Left | 2 |
| 3 | Leave | Right | 4 | Turn | Left | 7 |
| 4 | Leave | Right | 5 | Leave | Right | 4 |
| 5 | Leave | Right | 5 | Turn | Left | 6 |
| 6 | Leave | Left | 6 | Leave | Left | 2 |
| 7 | Turn | Left | 8 | — | — | — |
| 8 | Leave | Right | 9 | Leave | Right | 15 |
| 9 | Leave | Right | 10 | Leave | Right | 9 |
| 10 | Leave | Right | 11 | Leave | Right | 10 |
| 11 | Leave | Right | 11 | Turn | Left | 12 |
| 12 | Leave | Left | 12 | Leave | Left | 13 |
| 13 | Leave | Left | 14 | Leave | Left | 13 |
| 14 | Turn | Left | 8 | Leave | Left | 14 |
| 15 | Turn | Right | 16 | Turn | Right | 15 |
| 16 | Turn | Right | 17 | Turn | Right | 16 |
| 17 | Leave | Right | 17 | Turn | Right | 18 |
| 18 | — | — | — | Turn | None | — |

**Figure 11.5** The start of another game

**Figure 11.6** The finish of the game

This is an exercise in addition, which brings us to the heart of the matter. It can be shown that each of the fundamental operations of computing (addition, subtraction, multiplication, and division) can be performed by a suitably programmed Turing robot. The reader with a taste for such things may care to develop Table 11.2 into a program to perform a multiplication. All that is necessary is to count down a second row to the left of the first, and to perform the copying of the first row once for each head in the second; the details are mildly tedious, but the task is not difficult in principle. For another example, suppose the program in Table 11.3 to be altered so that the detection of a tail in state 18 causes the robot to reassume state 1 instead of stopping. If the initial configuration represents the numbers 1 and 1, the robot now computes the Fibonacci sequence 2,3,5,8,13,21,... (Figure 11.7).

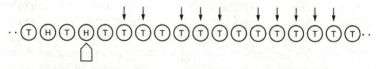

**Figure 11.7** The Fibonacci game. The arrowed tails become heads

But the operations of addition, subtraction, multiplication, and division are the building blocks from which all computations are constructed, and it follows that if a number can be computed at all then it can be computed by supplying a Turing robot with a suitable program. Furthermore, the playing of an automatic game is logically equivalent to the performance of a computation, so every automatic game can be simulated by a Turing robot. For example, it is possible to devise a representation of playing cards in terms of heads and tails, and then to program a Turing robot to play 'drive the old woman to bed'. It is perhaps unlikely that anyone would actually sit down and write the program, but the task is perfectly feasible.

# Turing's paradox

Having introduced Turing games, we can return to the question of deciding whether an automatic game can get into a loop.

We first observe that the action of a Turing robot is completely specified by a finite ordered array of numbers, and so can be represented by a coin arrangement containing a finite number of heads. 'Leave' and 'turn' can be represented by rows of one and two heads respectively; 'step left', 'step right', and 'stay put' can be represented similarly; and the number of the next state to be assumed can be represented by a row containing this number of heads. Figure 11.8 shows the start of a coin arrangement representing Table 11.2.

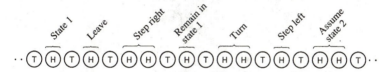

**Figure 11.8** The start of a coin representation of Table 11.2

Now we have remarked that every automatic game is equivalent to a Turing game, so the problem of deciding whether an automatic game terminates is equivalent to deciding whether a Turing game terminates. Let us therefore consider a general Turing program $P$ and a coin arrangement $c$ which contains a finite number of heads, let us assume that the robot starts at the rightmost head of $c$ (Figure 11.9), and let us postulate the existence of a Turing program $Q$ which will tell us whether $P$ now terminates. To fix our ideas, let us postulate this program $Q$ to be specified as follows: (i) the robot is to start at the rightmost head of $c$; (ii) the space to its right is to be occupied by the coin arrangement representing $P$ (Figure 11.10); (iii) the program $Q$ is always to terminate, the robot coming to rest on a head if $P$ terminates and on a tail if $P$ does not. It can be shown that this

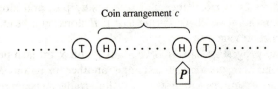

**Figure 11.9** The starting position for program $P$

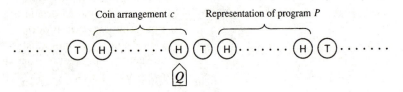

**Figure 11.10** The starting position for program $Q$

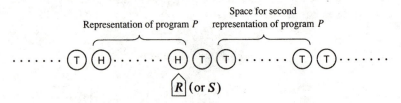

**Figure 11.11** The starting position for programs $R$ and $S$

detailed specification of $Q$ involves no loss of generality; if $Q$ exists at all, it can be specified in this form.

Now if such a program $Q$ does exist, we can develop from it a program $R$ which requires the coin array to contain only the arrangement representing $P$ (Figure 11.11) and which will detect whether $P$ terminates if the robot starts at the rightmost head of the coin arrangement representing $P$ itself. All that $R$ needs to do is to copy the coin arrangement representing $P$ into the space immediately to the right, to move the robot back to the starting position, and then to invoke $Q$.

Having obtained $R$, we can obtain a further program $S$ by replacing all the 'stop' commands within $R$ by commands to assume the following new state: 'seeing a head, leave it alone, stay put, and remain in this state; seeing a tail, leave it alone, stay put, and stop.' If $R$ stops on a tail, $S$ now does likewise; but if $R$ stops on a head, $S$ does not terminate.

Let us summarize all this. The programs $R$ and $S$ are each presented with a coin arrangement representing another program $P$. If $P$ terminates on being presented with the coin arrangement representing itself, $R$ stops on a head, and $S$ does not terminate; if $P$ does not terminate, $R$ stops on a tail, and $S$ stops likewise.

We now imagine $S$ presented with the coin arrangement representing itself (Figure 11.12), and we find its behaviour impossible to decide. If it terminates, then it does not; if it does not terminate, then it does. Hence no such program $S$ can exist, and so $R$ and $Q$ cannot exist either. In other words, there can be no general procedure for determining whether an automatic game always terminates.

Representation of program $S$

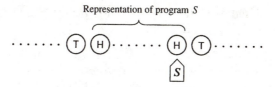

**Figure 11.12** The paradox

# The hole at the heart of mathematics

The essence of the technique in the previous section is quite simple. We establish a correspondence between the objects on which we are operating and the operations themselves, and we then show that a certain operation cannot exist, because it would generate a paradox if applied to the object corresponding to itself.

This technique can be applied more widely. A formal logical system is based on axioms and rules of construction, and a proposition is said to be 'provable' if it can be derived from the axioms by following the rules of construction. For a trivial example, let us suppose that we have an axiom 'Polly is a parrot' and a rule of construction which states that 'is a parrot' may be replaced by 'can talk'; then we can derive the statement 'Polly can talk' by applying the rule of construction to the axiom. To put matters more formally, let us suppose the phrases 'Polly', 'is a parrot', and 'can talk' to be denoted by $P$, $R$, and $K$

respectively, and the rule of construction 'throughout a given phrase $X$, replace $R$ by $K$' to be denoted by $C(X)$. If $PR$ is defined as an axiom, the symbol sequence '$C(PR)$' now becomes a formal proof of $PK$.

In a general logical system, every proposition $P$ has a complementary proposition $\sim P$ ('not $P$'), and this makes it necessary to refine the definition of provability. In principle, there are four possibilities.

(a) $P$ can be derived from the axioms by following the rules of construction, but $\sim P$ cannot. Only in this case is $P$ now said to be provable.

(b) $P$ cannot be derived from the axioms, but $\sim P$ can. $P$ is now said to be disprovable.

(c) Neither $P$ nor $\sim P$ can be derived from the axioms. $P$ is now said to be formally undecidable.

(d) Both $P$ and $\sim P$ can be derived from the axioms. In this case, the logical system is said to be inconsistent.

The belief of mathematicians, at least in respect of arithmetic, used to be that all propositions came into classes (a) and (b); proof or disproof always existed, even though it might not have been found. This belief was shattered in 1930, when Kurt Gödel discovered a proposition in class (c).[3]

What Gödel did was to assign a number to every symbol in a formal logical argument in such a way that the logical proposition 'The string of symbols $S$ constitutes a proof of the proposition $P$' became equivalent to an arithmetical proposition about the numbers representing $S$ and $P$. The proposition 'There is no string of symbols which constitutes a proof of proposition $P$' accordingly became equivalent to an arithmetical proposition about the number representing $P$; but, being itself a proposition, it was represented by a number, and when it was applied to the number representing itself, either a proof or a disproof was seen to lead to a contradiction. So this proposition was neither provable nor disprovable.

---

[3] The date of Gödel's discovery is variously reported as '1930' and '1931'. In fact he announced it to the Vienna Academy of Sciences in 1930, and I have adopted that date even though his detailed paper was not published until 1931. Gödel actually demonstrated his proposition to be undecidable only under the assumption that arithmetic is '$\omega$-consistent', a slightly more demanding requirement than ordinary consistency. The technical definition of $\omega$-consistency is beyond the scope of this book, and in any case it is now irrelevant; J. B. Rosser, using a more complicated proposition of the same type, reduced the requirement to ordinary consistency in 1936.

This was Gödel's first theorem, which was quite remarkable enough; but his second theorem was even more remarkable, because it identified the proposition 'Arithmetic is consistent' with the proposition 'There is no $P$ such that both $P$ and $\sim P$ are provable' and then showed the latter to be another undecidable proposition. So the consistency of arithmetic, on which the whole of mathematics relies, is formally unprovable.[4]

Gödel's theorems were known to Turing, and were indeed among the clues which led him to the discovery of his own paradox. With our modern knowledge of computers, however, it is easier to approach matters the other way round, and to use Turing's paradox to illuminate Gödel. That a program can be represented by a number is now obvious; this is how programs are stored in the memory of a computer. That a logical argument can be represented by a number is now perhaps just as obvious, but it was no means obvious in 1930.

In no sense has this been intended as a book of instruction; its sole object has been to entertain. Nevertheless, entertainment and instruction are more closely interwoven in mathematics than in any other subject, and few things demonstrate this more strikingly than the way in which speculations about the behaviour of elementary games for children can illuminate fundamental questions about the nature of reasoning itself.

---

[4] However, the fact that we cannot *prove* the consistency of arithmetic doesn't mean that it isn't *true*. 'Truth' and 'provability' are quite different things. Indeed, if the proposition 'Arithmetic is consistent' were *not* true, we *could* prove it, since if a logical system is inconsistent then any proposition which can be formulated in it can also be proved in it. Readers who would like to pursue this paradox further will find an excellent discussion in Raymond Smullyan's *Forever undecided: a puzzle guide to Gödel* (Knopf, 1987; Oxford, 1988).

# FURTHER READING

A writer on recreational mathematics is rather like a small boy in an orchard, or a tourist in an art gallery; he can take a nibble from each of several delights, or he can concentrate on one and ignore the rest. This book, unashamedly, has nibbled, and the books listed below form a natural next stage for readers whose appetites have been whetted by particular topics. New books are continually appearing, however, and this list should be supplemented by intelligent browsing in bookshops.

We start with two important general recommendations. All Martin Gardner's *Scientific American* books are worth reading; their range is wide, and their depth of treatment well judged. Not for nothing did Berlekamp, Conway, and Guy, in dedicating *Winning ways for your mathematical plays* to him, describe him as having brought 'more mathematics to more millions than anyone else'. In the same way, *Winning ways* itself (Academic Press, 1982) will long be a standard reference for games of pure skill. It contains an immense amount of material, a remarkably high proportion of which is original either with the authors themselves or with their students, colleagues, and friends. These books also contain extensive bibliographies which will assist readers who would like to pursue matters more deeply still.

Let us now take our various topics in turn. The odds affecting card and dice games, although important, are only a starting point, and among matters which you may wish to explore further are (i) the diagnosis of situations in which the usual probabilities appear not to apply (the literature of bridge, and in particular the many books by Terence Reese, being especially strong in this respect); (ii) card shuffling, which is discussed in Martin Gardner's *Mathematical carnival* (Knopf, 1975; George Allen and Unwin, 1978, subsequently reprinted by Penguin); and (iii) cheating and deception, which are perhaps best covered in *Scarne's complete guide to gambling* (Scarne, Simon and Schuster, 1961). Each of these takes matters beyond mere mathematics.

For example, Gardner's bibliography contains nearly twice as many items by magicians as by mathematicians, since it is very useful to a magician to know when an apparently random shuffle actually performs a known operation on a key card.

The published literature on the effect of chance on ball games is scattered among academic journals, and usually contains only crude analyses in which the variation due to external influences is not distinguished from the variation inherent in the game itself. This is perhaps a field in which you will now find it more profitable to perform an investigation yourself than to read about somebody else's. All that you need to do is to think about the game, to identify an apparently influential event, to analyse the effect of chance on this event, and to see to what extent this appears to explain the observed variation in the results. In some cases, you will find adequate data either in the newspapers or in standard annuals such as *Playfair* and *Wisden;* in others (the behaviour of tennis rallies, for example) you will have to collect your own. You should not be deterred from attempting such an investigation because you regard yourself as lacking statistical expertise. Some analyses of games do indeed require statistical subtlety, but the first stage in a competent statistical investigation is always an examination of the data using common sense, and in many cases this examination will tell you all that you need to know. If common sense fails to throw sufficient light, you can always ask an expert how to perform a more sophisticated analysis.

The estimation of skill from individual scores is an important topic within the study of statistics and economics, and those in search of further information should consult specialist textbooks. The problem here is not in finding material but in remembering amidst the mass of detail that *all* such estimates are subject to the limitations demonstrated in the early part of Chapter 5. The most widely available book on the estimation of skill at interactive games is currently Elo's *The rating of chessplayers, past and present* (Batsford, 1978), but this must be read with caution. Elo deserves the credit for being a pioneer and for doing a great deal of work, much of it before automatic computers were available to perform the arithmetic, but his work contains too many errors to be acceptable as a continuing standard.

The classic work on the von Neumann theory is *The theory of games and economic behaviour* (J. von Neumann and O. Morgenstern, Princeton, third edition 1953). It is not light reading, but it has no competitor. Chapters 2–4 cover two-person 'zero-sum' games, in which one player's gain is the other's loss, and the remaining two thirds of the book discuss more general games. Its major omission is of the use of computers to analyse games; the original 1944 edition was written before computers became available, and the later editions incorporate only minor revision. I know of no book for the layman which fills this gap, and indeed the task should not be attempted by other than a trained numerical analyst. If, as such, you wish to investigate further, your requirement is for a textbook whose title includes the words 'Theory of games' and 'Linear programming'; many such are in existence.

The literature of puzzles is vast. Most books on mathematical recreations contain some puzzle material; the Gardner books certainly do, as to a lesser extent does *Winning ways for your mathematical plays*. For that matter, three of the four books already published in the present series (*Sliding piece puzzles* by Hordern, *Rubik's cubic compendium* by Rubik and others, and my own *The ins and outs of peg solitaire*) are devoted to puzzles, and it will be interesting to see how the series develops in future. In general, however, the availability of puzzle books is so great that even a random browse in a games shop or bookshop has a reasonable chance of proving profitable.

The books for immediate further reading on games of pure skill are *Winning ways for your mathematical plays* and the earlier *On numbers and games* (Conway, Academic Press, 1976). There is appreciable overlap between them, but *On numbers and games* contains the more abstract mathematical material, whereas the larger *Winning ways for your mathematical plays* is oriented more towards the general reader. Because of the more formal mathematical basis of these books, their notation is different from ours, but the differences are soon absorbed. Our players Plus and Minus usually become 'Left' and 'Right'; our game $U$ becomes 'up', denoted by a small upward arrow; our worms become 'tiny' and 'miny'; and our infinite games $Z$, $A$, $LW$, $RW$, and $\infty$ become '$\omega$', '$1/\omega$', 'on', 'off', and 'dud'. As those who do read these books will discover, our Chapters 8–10 have touched only the outside of a very wide field.

The standard popular books on the topics of Chapter 11 are *Gödel, Escher, Bach: an eternal golden braid* by Douglas R. Hofstadter (Basic Books/Harvester 1979, subsequently reprinted by Penguin) and the much more recent *Forever undecided: a puzzle guide to Gödel* by Raymond Smullyan (Knopf, 1987; Oxford, 1988). In so far as they overlap, Smullyan seems to me greatly superior, but the Turing material is only in Hofstadter. Several translations of Gödel's actual paper also exist in English (for example, *On formally undecidable propositions of Principia Mathematica and related systems*, translated by B. Meltzer with an introduction by R. B. Braithwaite, Oliver and Boyd, 1962), but they are not easy reading for anybody unused to the notation of formal logic.

There are also periodicals: magazines, newspaper columns, and so on. These provide a rich source of material, but their average lifetime is so short that there is little point even in listing those which are currently prominent; all too many will have vanished within a few years, while others will have arisen to take their place. Nearly every successful games magazine or newspaper column depends on the energy and ability of one man, and volatility is inevitable.

There are in fact six main classes.

(a) Magazines produced as private hobbies, the editor having a separate full-time job. These are the most volatile of all. If the editor thrives, the magazine thrives with him; if he falls ill, or finds that the regular editorial grind has become greater than he can accommodate, the magazine lapses. Yet many of the best magazines have come into this class. T. R. Dawson's legendary *Fairy chess review* was one, and all who have trodden this path since have marvelled at the amount of work which he did.

(b) Journals published by societies whose members are interested in a particular game. These have much in common with the previous class, since the editor and his contributors are usually unpaid hobbyists, but the broader base of a typical society means that they are slightly less volatile.

(c) Journals published for teachers of mathematics. Many of these contain excellent puzzle columns, and some contain occasional analyses of games (though the emphasis is usually on the mathematical lessons that can be learned rather than on the behaviour of the game itself).

(d) Magazines produced by general commercial publishers. Such a publisher usually produces a range of magazines, so that the overheads of distribution and administration can be shared. Even so, each magazine must generate enough income to pay its editor and its contributors, to cover its other costs, and to contribute its share to the overall profits, so the 'logic' of the market-place is always present. Puzzles follow fashions, as Rubik's cube has graphically demonstrated. While a craze lasts, a magazine can adapt its content to it, and sales rise. When the craze has worked itself out, the magazine must find other material, and the result may not be commercially viable.

(e) Magazines produced by specialist commercial organizations. Typically, the magazine is devoted to a particular game, and the publisher also acts as a mail-order retailer of books and equipment. As long as the game can support such a retailer, this is a good recipe for stability; the administrative overheads are shared among more than one activity, and the magazine provides both a focal point and a highly efficient advertising medium. Yet personal volatility may intrude even here. The corporate front of such an organization often conceals an active involvement of only one or two people, and illness or advancing age may cause problems.

(f) Columns in newspapers and general journals. Although these are frequently long-lasting, their commercial purpose is to boost the sales of the parent publication, and they are always subject to the changing views of proprietors as to what potential readers really want. A good column also reflects the individual knowledge and ability of its writer, to an extent which may be realized only after his departure. For many years, Martin Gardner's column *Mathematical games* in *Scientific American* was one of the fixed points in the recreational mathematician's universe; yet on his retirement, the column became *Computer recreations*, and changed noticeably in character.

In this opaque situation, by far the best guide is personal recommendation. Failing this, a column in a teaching journal, newspaper, or general magazine makes a good starting point. Not only are the best of such columns very good in themselves, but some columnists make a point of regularly mentioning magazines which they believe worthy of notice. It is also a reasonable strategy to make occasional random purchases of the puzzle magazines which are sold on bookstalls at railway stations and airports. These vary widely in quality, but even the more feeble contain an occasional nugget among the dross, and at the worst you have only wasted the cost of one issue.

It may seem depressing that no better advice is possible, yet this is in the nature of the subject. One of the beauties of recreational mathematics is that it is an individual pursuit; it does not demand elaborate organization or expensive equipment, merely interest and ability. The volatility of its magazines is an inevitable consequence.

# INDEX

# A CATALOG OF SELECTED
# DOVER BOOKS
## IN SCIENCE AND MATHEMATICS

# Astronomy

BURNHAM'S CELESTIAL HANDBOOK, Robert Burnham, Jr. Thorough guide to the stars beyond our solar system. Exhaustive treatment. Alphabetical by constellation: Andromeda to Cetus in Vol. 1; Chamaeleon to Orion in Vol. 2; and Pavo to Vulpecula in Vol. 3. Hundreds of illustrations. Index in Vol. 3. 2,000pp. 6⅛ x 9¼.
Vol. I: 0-486-23567-X
Vol. II: 0-486-23568-8
Vol. III: 0-486-23673-0

EXPLORING THE MOON THROUGH BINOCULARS AND SMALL TELESCOPES, Ernest H. Cherrington, Jr. Informative, profusely illustrated guide to locating and identifying craters, rills, seas, mountains, other lunar features. Newly revised and updated with special section of new photos. Over 100 photos and diagrams. 240pp. 8¼ x 11. 0-486-24491-1

THE EXTRATERRESTRIAL LIFE DEBATE, 1750–1900, Michael J. Crowe. First detailed, scholarly study in English of the many ideas that developed from 1750 to 1900 regarding the existence of intelligent extraterrestrial life. Examines ideas of Kant, Herschel, Voltaire, Percival Lowell, many other scientists and thinkers. 16 illustrations. 704pp. 5⅜ x 8½. 0-486-40675-X

THEORIES OF THE WORLD FROM ANTIQUITY TO THE COPERNICAN REVOLUTION, Michael J. Crowe. Newly revised edition of an accessible, enlightening book recreates the change from an earth-centered to a sun-centered conception of the solar system. 242pp. 5⅜ x 8½. 0-486-41444-2

A HISTORY OF ASTRONOMY, A. Pannekoek. Well-balanced, carefully reasoned study covers such topics as Ptolemaic theory, work of Copernicus, Kepler, Newton, Eddington's work on stars, much more. Illustrated. References. 521pp. 5⅜ x 8½.
0-486-65994-1

A COMPLETE MANUAL OF AMATEUR ASTRONOMY: TOOLS AND TECHNIQUES FOR ASTRONOMICAL OBSERVATIONS, P. Clay Sherrod with Thomas L. Koed. Concise, highly readable book discusses: selecting, setting up and maintaining a telescope; amateur studies of the sun; lunar topography and occultations; observations of Mars, Jupiter, Saturn, the minor planets and the stars; an introduction to photoelectric photometry; more. 1981 ed. 124 figures. 25 halftones. 37 tables. 335pp. 6½ x 9¼. 0-486-40675-X

AMATEUR ASTRONOMER'S HANDBOOK, J. B. Sidgwick. Timeless, comprehensive coverage of telescopes, mirrors, lenses, mountings, telescope drives, micrometers, spectroscopes, more. 189 illustrations. 576pp. 5⅝ x 8¼. (Available in U.S. only.)
0-486-24034-7

STARS AND RELATIVITY, Ya. B. Zel'dovich and I. D. Novikov. Vol. 1 of *Relativistic Astrophysics* by famed Russian scientists. General relativity, properties of matter under astrophysical conditions, stars, and stellar systems. Deep physical insights, clear presentation. 1971 edition. References. 544pp. 5⅝ x 8¼. 0-486-69424-0

# Chemistry

THE SCEPTICAL CHYMIST: THE CLASSIC 1661 TEXT, Robert Boyle. Boyle defines the term "element," asserting that all natural phenomena can be explained by the motion and organization of primary particles. 1911 ed. viii+232pp. 5⅜ x 8½.
0-486-42825-7

RADIOACTIVE SUBSTANCES, Marie Curie. Here is the celebrated scientist's doctoral thesis, the prelude to her receipt of the 1903 Nobel Prize. Curie discusses establishing atomic character of radioactivity found in compounds of uranium and thorium; extraction from pitchblende of polonium and radium; isolation of pure radium chloride; determination of atomic weight of radium; plus electric, photographic, luminous, heat, color effects of radioactivity. ii+94pp. 5⅜ x 8½. 0-486-42550-9

CHEMICAL MAGIC, Leonard A. Ford. Second Edition, Revised by E. Winston Grundmeier. Over 100 unusual stunts demonstrating cold fire, dust explosions, much more. Text explains scientific principles and stresses safety precautions. 128pp. 5⅜ x 8½. 0-486-67628-5

THE DEVELOPMENT OF MODERN CHEMISTRY, Aaron J. Ihde. Authoritative history of chemistry from ancient Greek theory to 20th-century innovation. Covers major chemists and their discoveries. 209 illustrations. 14 tables. Bibliographies. Indices. Appendices. 851pp. 5⅜ x 8½. 0-486-64235-6

CATALYSIS IN CHEMISTRY AND ENZYMOLOGY, William P. Jencks. Exceptionally clear coverage of mechanisms for catalysis, forces in aqueous solution, carbonyl- and acyl-group reactions, practical kinetics, more. 864pp. 5⅜ x 8½.
0-486-65460-5

ELEMENTS OF CHEMISTRY, Antoine Lavoisier. Monumental classic by founder of modern chemistry in remarkable reprint of rare 1790 Kerr translation. A must for every student of chemistry or the history of science. 539pp. 5⅜ x 8½. 0-486-64624-6

THE HISTORICAL BACKGROUND OF CHEMISTRY, Henry M. Leicester. Evolution of ideas, not individual biography. Concentrates on formulation of a coherent set of chemical laws. 260pp. 5⅜ x 8½. 0-486-61053-5

A SHORT HISTORY OF CHEMISTRY, J. R. Partington. Classic exposition explores origins of chemistry, alchemy, early medical chemistry, nature of atmosphere, theory of valency, laws and structure of atomic theory, much more. 428pp. 5⅜ x 8½. (Available in U.S. only.) 0-486-65977-1

GENERAL CHEMISTRY, Linus Pauling. Revised 3rd edition of classic first-year text by Nobel laureate. Atomic and molecular structure, quantum mechanics, statistical mechanics, thermodynamics correlated with descriptive chemistry. Problems. 992pp. 5⅜ x 8½. 0-486-65622-5

FROM ALCHEMY TO CHEMISTRY, John Read. Broad, humanistic treatment focuses on great figures of chemistry and ideas that revolutionized the science. 50 illustrations. 240pp. 5⅜ x 8½. 0-486-28690-8

# Engineering

DE RE METALLICA, Georgius Agricola. The famous Hoover translation of greatest treatise on technological chemistry, engineering, geology, mining of early modern times (1556). All 289 original woodcuts. 638pp. 6¾ x 11.          0-486-60006-8

FUNDAMENTALS OF ASTRODYNAMICS, Roger Bate et al. Modern approach developed by U.S. Air Force Academy. Designed as a first course. Problems, exercises. Numerous illustrations. 455pp. 5⅜ x 8½.          0-486-60061-0

DYNAMICS OF FLUIDS IN POROUS MEDIA, Jacob Bear. For advanced students of ground water hydrology, soil mechanics and physics, drainage and irrigation engineering and more. 335 illustrations. Exercises, with answers. 784pp. 6⅛ x 9¼.
0-486-65675-6

THEORY OF VISCOELASTICITY (Second Edition), Richard M. Christensen. Complete consistent description of the linear theory of the viscoelastic behavior of materials. Problem-solving techniques discussed. 1982 edition. 29 figures. xiv+364pp. 6⅛ x 9¼.          0-486-42880-X

MECHANICS, J. P. Den Hartog. A classic introductory text or refresher. Hundreds of applications and design problems illuminate fundamentals of trusses, loaded beams and cables, etc. 334 answered problems. 462pp. 5⅜ x 8½.          0-486-60754-2

MECHANICAL VIBRATIONS, J. P. Den Hartog. Classic textbook offers lucid explanations and illustrative models, applying theories of vibrations to a variety of practical industrial engineering problems. Numerous figures. 233 problems, solutions. Appendix. Index. Preface. 436pp. 5⅜ x 8½.          0-486-64785-4

STRENGTH OF MATERIALS, J. P. Den Hartog. Full, clear treatment of basic material (tension, torsion, bending, etc.) plus advanced material on engineering methods, applications. 350 answered problems. 323pp. 5⅜ x 8½.          0-486-60755-0

A HISTORY OF MECHANICS, René Dugas. Monumental study of mechanical principles from antiquity to quantum mechanics. Contributions of ancient Greeks, Galileo, Leonardo, Kepler, Lagrange, many others. 671pp. 5⅜ x 8½.  0-486-65632-2

STABILITY THEORY AND ITS APPLICATIONS TO STRUCTURAL MECHANICS, Clive L. Dym. Self-contained text focuses on Koiter postbuckling analyses, with mathematical notions of stability of motion. Basing minimum energy principles for static stability upon dynamic concepts of stability of motion, it develops asymptotic buckling and postbuckling analyses from potential energy considerations, with applications to columns, plates, and arches. 1974 ed. 208pp. 5⅜ x 8½.
0-486-42541-X

METAL FATIGUE, N. E. Frost, K. J. Marsh, and L. P. Pook. Definitive, clearly written, and well-illustrated volume addresses all aspects of the subject, from the historical development of understanding metal fatigue to vital concepts of the cyclic stress that causes a crack to grow. Includes 7 appendixes. 544pp. 5⅜ x 8½.  0-486-40927-9

ROCKETS, Robert Goddard. Two of the most significant publications in the history of rocketry and jet propulsion: "A Method of Reaching Extreme Altitudes" (1919) and "Liquid Propellant Rocket Development" (1936). 128pp. 5⅜ x 8½.  0-486-42537-1

STATISTICAL MECHANICS: PRINCIPLES AND APPLICATIONS, Terrell L. Hill. Standard text covers fundamentals of statistical mechanics, applications to fluctuation theory, imperfect gases, distribution functions, more. 448pp. 5⅜ x 8½.
0-486-65390-0

ENGINEERING AND TECHNOLOGY 1650–1750: ILLUSTRATIONS AND TEXTS FROM ORIGINAL SOURCES, Martin Jensen. Highly readable text with more than 200 contemporary drawings and detailed engravings of engineering projects dealing with surveying, leveling, materials, hand tools, lifting equipment, transport and erection, piling, bailing, water supply, hydraulic engineering, and more. Among the specific projects outlined-transporting a 50-ton stone to the Louvre, erecting an obelisk, building timber locks, and dredging canals. 207pp. 8⅜ x 11¼.
0-486-42232-1

THE VARIATIONAL PRINCIPLES OF MECHANICS, Cornelius Lanczos. Graduate level coverage of calculus of variations, equations of motion, relativistic mechanics, more. First inexpensive paperbound edition of classic treatise. Index. Bibliography. 418pp. 5⅜ x 8½.  0-486-65067-7

PROTECTION OF ELECTRONIC CIRCUITS FROM OVERVOLTAGES, Ronald B. Standler. Five-part treatment presents practical rules and strategies for circuits designed to protect electronic systems from damage by transient overvoltages. 1989 ed. xxiv+434pp. 6⅛ x 9¼.  0-486-42552-5

ROTARY WING AERODYNAMICS, W. Z. Stepniewski. Clear, concise text covers aerodynamic phenomena of the rotor and offers guidelines for helicopter performance evaluation. Originally prepared for NASA. 537 figures. 640pp. 6⅛ x 9¼.
0-486-64647-5

INTRODUCTION TO SPACE DYNAMICS, William Tyrrell Thomson. Comprehensive, classic introduction to space-flight engineering for advanced undergraduate and graduate students. Includes vector algebra, kinematics, transformation of coordinates. Bibliography. Index. 352pp. 5⅜ x 8½.  0-486-65113-4

HISTORY OF STRENGTH OF MATERIALS, Stephen P. Timoshenko. Excellent historical survey of the strength of materials with many references to the theories of elasticity and structure. 245 figures. 452pp. 5⅜ x 8½.  0-486-61187-6

ANALYTICAL FRACTURE MECHANICS, David J. Unger. Self-contained text supplements standard fracture mechanics texts by focusing on analytical methods for determining crack-tip stress and strain fields. 336pp. 6⅛ x 9¼.  0-486-41737-9

STATISTICAL MECHANICS OF ELASTICITY, J. H. Weiner. Advanced, self-contained treatment illustrates general principles and elastic behavior of solids. Part 1, based on classical mechanics, studies thermoelastic behavior of crystalline and polymeric solids. Part 2, based on quantum mechanics, focuses on interatomic force laws, behavior of solids, and thermally activated processes. For students of physics and chemistry and for polymer physicists. 1983 ed. 96 figures. 496pp. 5⅜ x 8½.
0-486-42260-7

# Mathematics

FUNCTIONAL ANALYSIS (Second Corrected Edition), George Bachman and Lawrence Narici. Excellent treatment of subject geared toward students with background in linear algebra, advanced calculus, physics and engineering. Text covers introduction to inner-product spaces, normed, metric spaces, and topological spaces; complete orthonormal sets, the Hahn-Banach Theorem and its consequences, and many other related subjects. 1966 ed. 544pp. 6⅛ x 9¼. 0-486-40251-7

ASYMPTOTIC EXPANSIONS OF INTEGRALS, Norman Bleistein & Richard A. Handelsman. Best introduction to important field with applications in a variety of scientific disciplines. New preface. Problems. Diagrams. Tables. Bibliography. Index. 448pp. 5⅜ x 8½. 0-486-65082-0

VECTOR AND TENSOR ANALYSIS WITH APPLICATIONS, A. I. Borisenko and I. E. Tarapov. Concise introduction. Worked-out problems, solutions, exercises. 257pp. 5⅝ x 8¼. 0-486-63833-2

AN INTRODUCTION TO ORDINARY DIFFERENTIAL EQUATIONS, Earl A. Coddington. A thorough and systematic first course in elementary differential equations for undergraduates in mathematics and science, with many exercises and problems (with answers). Index. 304pp. 5⅜ x 8½. 0-486-65942-9

FOURIER SERIES AND ORTHOGONAL FUNCTIONS, Harry F. Davis. An incisive text combining theory and practical example to introduce Fourier series, orthogonal functions and applications of the Fourier method to boundary-value problems. 570 exercises. Answers and notes. 416pp. 5⅜ x 8½. 0-486-65973-9

COMPUTABILITY AND UNSOLVABILITY, Martin Davis. Classic graduate-level introduction to theory of computability, usually referred to as theory of recurrent functions. New preface and appendix. 288pp. 5⅜ x 8½. 0-486-61471-9

ASYMPTOTIC METHODS IN ANALYSIS, N. G. de Bruijn. An inexpensive, comprehensive guide to asymptotic methods—the pioneering work that teaches by explaining worked examples in detail. Index. 224pp. 5⅜ x 8½ 0-486-64221-6

APPLIED COMPLEX VARIABLES, John W. Dettman. Step-by-step coverage of fundamentals of analytic function theory—plus lucid exposition of five important applications: Potential Theory; Ordinary Differential Equations; Fourier Transforms; Laplace Transforms; Asymptotic Expansions. 66 figures. Exercises at chapter ends. 512pp. 5⅜ x 8½. 0-486-64670-X

INTRODUCTION TO LINEAR ALGEBRA AND DIFFERENTIAL EQUATIONS, John W. Dettman. Excellent text covers complex numbers, determinants, orthonormal bases, Laplace transforms, much more. Exercises with solutions. Undergraduate level. 416pp. 5⅜ x 8½. 0-486-65191-6

RIEMANN'S ZETA FUNCTION, H. M. Edwards. Superb, high-level study of landmark 1859 publication entitled "On the Number of Primes Less Than a Given Magnitude" traces developments in mathematical theory that it inspired. xiv+315pp. 5⅜ x 8½. 0-486-41740-9

CALCULUS OF VARIATIONS WITH APPLICATIONS, George M. Ewing. Applications-oriented introduction to variational theory develops insight and promotes understanding of specialized books, research papers. Suitable for advanced undergraduate/graduate students as primary, supplementary text. 352pp. 5⅜ x 8½. 0-486-64856-7

COMPLEX VARIABLES, Francis J. Flanigan. Unusual approach, delaying complex algebra till harmonic functions have been analyzed from real variable viewpoint. Includes problems with answers. 364pp. 5⅜ x 8½. 0-486-61388-7

AN INTRODUCTION TO THE CALCULUS OF VARIATIONS, Charles Fox. Graduate-level text covers variations of an integral, isoperimetrical problems, least action, special relativity, approximations, more. References. 279pp. 5⅜ x 8½. 0-486-65499-0

COUNTEREXAMPLES IN ANALYSIS, Bernard R. Gelbaum and John M. H. Olmsted. These counterexamples deal mostly with the part of analysis known as "real variables." The first half covers the real number system, and the second half encompasses higher dimensions. 1962 edition. xxiv+198pp. 5⅜ x 8½. 0-486-42875-3

CATASTROPHE THEORY FOR SCIENTISTS AND ENGINEERS, Robert Gilmore. Advanced-level treatment describes mathematics of theory grounded in the work of Poincaré, R. Thom, other mathematicians. Also important applications to problems in mathematics, physics, chemistry and engineering. 1981 edition. References. 28 tables. 397 black-and-white illustrations. xvii + 666pp. 6⅛ x 9¼. 0-486-67539-4

INTRODUCTION TO DIFFERENCE EQUATIONS, Samuel Goldberg. Exceptionally clear exposition of important discipline with applications to sociology, psychology, economics. Many illustrative examples; over 250 problems. 260pp. 5⅜ x 8½. 0-486-65084-7

NUMERICAL METHODS FOR SCIENTISTS AND ENGINEERS, Richard Hamming. Classic text stresses frequency approach in coverage of algorithms, polynomial approximation, Fourier approximation, exponential approximation, other topics. Revised and enlarged 2nd edition. 721pp. 5⅜ x 8½. 0-486-65241-6

INTRODUCTION TO NUMERICAL ANALYSIS (2nd Edition), F. B. Hildebrand. Classic, fundamental treatment covers computation, approximation, interpolation, numerical differentiation and integration, other topics. 150 new problems. 669pp. 5⅜ x 8½. 0-486-65363-3

THREE PEARLS OF NUMBER THEORY, A. Y. Khinchin. Three compelling puzzles require proof of a basic law governing the world of numbers. Challenges concern van der Waerden's theorem, the Landau-Schnirelmann hypothesis and Mann's theorem, and a solution to Waring's problem. Solutions included. 64pp. 5⅜ x 8½. 0-486-40026-3

THE PHILOSOPHY OF MATHEMATICS: AN INTRODUCTORY ESSAY, Stephan Körner. Surveys the views of Plato, Aristotle, Leibniz & Kant concerning propositions and theories of applied and pure mathematics. Introduction. Two appendices. Index. 198pp. 5⅜ x 8½. 0-486-25048-2

INTRODUCTORY REAL ANALYSIS, A.N. Kolmogorov, S. V. Fomin. Translated by Richard A. Silverman. Self-contained, evenly paced introduction to real and functional analysis. Some 350 problems. 403pp. 5⅜ x 8½. 0-486-61226-0

APPLIED ANALYSIS, Cornelius Lanczos. Classic work on analysis and design of finite processes for approximating solution of analytical problems. Algebraic equations, matrices, harmonic analysis, quadrature methods, much more. 559pp. 5⅜ x 8½. 0-486-65656-X

AN INTRODUCTION TO ALGEBRAIC STRUCTURES, Joseph Landin. Superb self-contained text covers "abstract algebra": sets and numbers, theory of groups, theory of rings, much more. Numerous well-chosen examples, exercises. 247pp. 5⅜ x 8½. 0-486-65940-2

QUALITATIVE THEORY OF DIFFERENTIAL EQUATIONS, V. V. Nemytskii and V.V. Stepanov. Classic graduate-level text by two prominent Soviet mathematicians covers classical differential equations as well as topological dynamics and ergodic theory. Bibliographies. 523pp. 5⅜ x 8½. 0-486-65954-2

THEORY OF MATRICES, Sam Perlis. Outstanding text covering rank, nonsingularity and inverses in connection with the development of canonical matrices under the relation of equivalence, and without the intervention of determinants. Includes exercises. 237pp. 5⅜ x 8½. 0-486-66810-X

INTRODUCTION TO ANALYSIS, Maxwell Rosenlicht. Unusually clear, accessible coverage of set theory, real number system, metric spaces, continuous functions, Riemann integration, multiple integrals, more. Wide range of problems. Undergraduate level. Bibliography. 254pp. 5⅜ x 8½. 0-486-65038-3

MODERN NONLINEAR EQUATIONS, Thomas L. Saaty. Emphasizes practical solution of problems; covers seven types of equations. ". . . a welcome contribution to the existing literature...."–*Math Reviews*. 490pp. 5⅜ x 8½. 0-486-64232-1

MATRICES AND LINEAR ALGEBRA, Hans Schneider and George Phillip Barker. Basic textbook covers theory of matrices and its applications to systems of linear equations and related topics such as determinants, eigenvalues and differential equations. Numerous exercises. 432pp. 5⅜ x 8½. 0-486-66014-1

LINEAR ALGEBRA, Georgi E. Shilov. Determinants, linear spaces, matrix algebras, similar topics. For advanced undergraduates, graduates. Silverman translation. 387pp. 5⅜ x 8½. 0-486-63518-X

ELEMENTS OF REAL ANALYSIS, David A. Sprecher. Classic text covers fundamental concepts, real number system, point sets, functions of a real variable, Fourier series, much more. Over 500 exercises. 352pp. 5⅜ x 8½. 0-486-65385-4

SET THEORY AND LOGIC, Robert R. Stoll. Lucid introduction to unified theory of mathematical concepts. Set theory and logic seen as tools for conceptual understanding of real number system. 496pp. 5⅜ x 8¼. 0-486-63829-4

TENSOR CALCULUS, J.L. Synge and A. Schild. Widely used introductory text covers spaces and tensors, basic operations in Riemannian space, non-Riemannian spaces, etc. 324pp. 5⅜ x 8¼. 0-486-63612-7

ORDINARY DIFFERENTIAL EQUATIONS, Morris Tenenbaum and Harry Pollard. Exhaustive survey of ordinary differential equations for undergraduates in mathematics, engineering, science. Thorough analysis of theorems. Diagrams. Bibliography. Index. 818pp. 5⅜ x 8½. 0-486-64940-7

INTEGRAL EQUATIONS, F. G. Tricomi. Authoritative, well-written treatment of extremely useful mathematical tool with wide applications. Volterra Equations, Fredholm Equations, much more. Advanced undergraduate to graduate level. Exercises. Bibliography. 238pp. 5⅜ x 8½. 0-486-64828-1

FOURIER SERIES, Georgi P. Tolstov. Translated by Richard A. Silverman. A valuable addition to the literature on the subject, moving clearly from subject to subject and theorem to theorem. 107 problems, answers. 336pp. 5⅜ x 8½. 0-486-63317-9

INTRODUCTION TO MATHEMATICAL THINKING, Friedrich Waismann. Examinations of arithmetic, geometry, and theory of integers; rational and natural numbers; complete induction; limit and point of accumulation; remarkable curves; complex and hypercomplex numbers, more. 1959 ed. 27 figures. xii+260pp. 5⅜ x 8½. 0-486-63317-9

POPULAR LECTURES ON MATHEMATICAL LOGIC, Hao Wang. Noted logician's lucid treatment of historical developments, set theory, model theory, recursion theory and constructivism, proof theory, more. 3 appendixes. Bibliography. 1981 edition. ix + 283pp. 5⅜ x 8½. 0-486-67632-3

CALCULUS OF VARIATIONS, Robert Weinstock. Basic introduction covering isoperimetric problems, theory of elasticity, quantum mechanics, electrostatics, etc. Exercises throughout. 326pp. 5⅜ x 8½. 0-486-63069-2

THE CONTINUUM: A CRITICAL EXAMINATION OF THE FOUNDATION OF ANALYSIS, Hermann Weyl. Classic of 20th-century foundational research deals with the conceptual problem posed by the continuum. 156pp. 5⅜ x 8½. 0-486-67982-9

CHALLENGING MATHEMATICAL PROBLEMS WITH ELEMENTARY SOLUTIONS, A. M. Yaglom and I. M. Yaglom. Over 170 challenging problems on probability theory, combinatorial analysis, points and lines, topology, convex polygons, many other topics. Solutions. Total of 445pp. 5⅜ x 8½. Two-vol. set. Vol. I: 0-486-65536-9 Vol. II: 0-486-65537-7

INTRODUCTION TO PARTIAL DIFFERENTIAL EQUATIONS WITH APPLICATIONS, E. C. Zachmanoglou and Dale W. Thoe. Essentials of partial differential equations applied to common problems in engineering and the physical sciences. Problems and answers. 416pp. 5⅜ x 8½. 0-486-65251-3

THE THEORY OF GROUPS, Hans J. Zassenhaus. Well-written graduate-level text acquaints reader with group-theoretic methods and demonstrates their usefulness in mathematics. Axioms, the calculus of complexes, homomorphic mapping, *p*-group theory, more. 276pp. 5⅜ x 8½. 0-486-40922-8

# Math–Decision Theory, Statistics, Probability

ELEMENTARY DECISION THEORY, Herman Chernoff and Lincoln E. Moses. Clear introduction to statistics and statistical theory covers data processing, probability and random variables, testing hypotheses, much more. Exercises. 364pp. 5⅜ x 8½. 0-486-65218-1

STATISTICS MANUAL, Edwin L. Crow et al. Comprehensive, practical collection of classical and modern methods prepared by U.S. Naval Ordnance Test Station. Stress on use. Basics of statistics assumed. 288pp. 5⅜ x 8½. 0-486-60599-X

SOME THEORY OF SAMPLING, William Edwards Deming. Analysis of the problems, theory and design of sampling techniques for social scientists, industrial managers and others who find statistics important at work. 61 tables. 90 figures. xvii +602pp. 5⅜ x 8½. 0-486-64684-X

LINEAR PROGRAMMING AND ECONOMIC ANALYSIS, Robert Dorfman, Paul A. Samuelson and Robert M. Solow. First comprehensive treatment of linear programming in standard economic analysis. Game theory, modern welfare economics, Leontief input-output, more. 525pp. 5⅜ x 8½. 0-486-65491-5

PROBABILITY: AN INTRODUCTION, Samuel Goldberg. Excellent basic text covers set theory, probability theory for finite sample spaces, binomial theorem, much more. 360 problems. Bibliographies. 322pp. 5⅜ x 8½. 0-486-65252-1

GAMES AND DECISIONS: INTRODUCTION AND CRITICAL SURVEY, R. Duncan Luce and Howard Raiffa. Superb nontechnical introduction to game theory, primarily applied to social sciences. Utility theory, zero-sum games, n-person games, decision-making, much more. Bibliography. 509pp. 5⅜ x 8½. 0-486-65943-7

INTRODUCTION TO THE THEORY OF GAMES, J. C. C. McKinsey. This comprehensive overview of the mathematical theory of games illustrates applications to situations involving conflicts of interest, including economic, social, political, and military contexts. Appropriate for advanced undergraduate and graduate courses; advanced calculus a prerequisite. 1952 ed. x+372pp. 5⅜ x 8½. 0-486-42811-7

FIFTY CHALLENGING PROBLEMS IN PROBABILITY WITH SOLUTIONS, Frederick Mosteller. Remarkable puzzlers, graded in difficulty, illustrate elementary and advanced aspects of probability. Detailed solutions. 88pp. 5⅜ x 8½. 65355-2

PROBABILITY THEORY: A CONCISE COURSE, Y. A. Rozanov. Highly readable, self-contained introduction covers combination of events, dependent events, Bernoulli trials, etc. 148pp. 5⅜ x 8¼. 0-486-63544-9

STATISTICAL METHOD FROM THE VIEWPOINT OF QUALITY CONTROL, Walter A. Shewhart. Important text explains regulation of variables, uses of statistical control to achieve quality control in industry, agriculture, other areas. 192pp. 5⅜ x 8½. 0-486-65232-7

## Math–Geometry and Topology

ELEMENTARY CONCEPTS OF TOPOLOGY, Paul Alexandroff. Elegant, intuitive approach to topology from set-theoretic topology to Betti groups; how concepts of topology are useful in math and physics. 25 figures. 57pp. 5⅜ x 8½. 0-486-60747-X

COMBINATORIAL TOPOLOGY, P. S. Alexandrov. Clearly written, well-organized, three-part text begins by dealing with certain classic problems without using the formal techniques of homology theory and advances to the central concept, the Betti groups. Numerous detailed examples. 654pp. 5⅜ x 8½. 0-486-40179-0

EXPERIMENTS IN TOPOLOGY, Stephen Barr. Classic, lively explanation of one of the byways of mathematics. Klein bottles, Moebius strips, projective planes, map coloring, problem of the Koenigsberg bridges, much more, described with clarity and wit. 43 figures. 210pp. 5⅜ x 8½. 0-486-25933-1

THE GEOMETRY OF RENÉ DESCARTES, René Descartes. The great work founded analytical geometry. Original French text, Descartes's own diagrams, together with definitive Smith-Latham translation. 244pp. 5⅜ x 8½. 0-486-60068-8

EUCLIDEAN GEOMETRY AND TRANSFORMATIONS, Clayton W. Dodge. This introduction to Euclidean geometry emphasizes transformations, particularly isometries and similarities. Suitable for undergraduate courses, it includes numerous examples, many with detailed answers. 1972 ed. viii+296pp. 6⅛ x 9¼. 0-486-43476-1

PRACTICAL CONIC SECTIONS: THE GEOMETRIC PROPERTIES OF ELLIPSES, PARABOLAS AND HYPERBOLAS, J. W. Downs. This text shows how to create ellipses, parabolas, and hyperbolas. It also presents historical background on their ancient origins and describes the reflective properties and roles of curves in design applications. 1993 ed. 98 figures. xii+100pp. 6½ x 9¼. 0-486-42876-1

THE THIRTEEN BOOKS OF EUCLID'S ELEMENTS, translated with introduction and commentary by Sir Thomas L. Heath. Definitive edition. Textual and linguistic notes, mathematical analysis. 2,500 years of critical commentary. Unabridged. 1,414pp. 5⅜ x 8½. Three-vol. set.
Vol. I: 0-486-60088-2 Vol. II: 0-486-60089-0 Vol. III: 0-486-60090-4

SPACE AND GEOMETRY: IN THE LIGHT OF PHYSIOLOGICAL, PSYCHOLOGICAL AND PHYSICAL INQUIRY, Ernst Mach. Three essays by an eminent philosopher and scientist explore the nature, origin, and development of our concepts of space, with a distinctness and precision suitable for undergraduate students and other readers. 1906 ed. vi+148pp. 5⅜ x 8½. 0-486-43909-7

GEOMETRY OF COMPLEX NUMBERS, Hans Schwerdtfeger. Illuminating, widely praised book on analytic geometry of circles, the Moebius transformation, and two-dimensional non-Euclidean geometries. 200pp. 5⅝ x 8¼. 0-486-63830-8

DIFFERENTIAL GEOMETRY, Heinrich W. Guggenheimer. Local differential geometry as an application of advanced calculus and linear algebra. Curvature, transformation groups, surfaces, more. Exercises. 62 figures. 378pp. 5⅜ x 8½. 0-486-63433-7

# History of Math

THE WORKS OF ARCHIMEDES, Archimedes (T. L. Heath, ed.). Topics include the famous problems of the ratio of the areas of a cylinder and an inscribed sphere; the measurement of a circle; the properties of conoids, spheroids, and spirals; and the quadrature of the parabola. Informative introduction. clxxxvi+326pp. 5⅜ x 8½.
0-486-42084-1

A SHORT ACCOUNT OF THE HISTORY OF MATHEMATICS, W. W. Rouse Ball. One of clearest, most authoritative surveys from the Egyptians and Phoenicians through 19th-century figures such as Grassman, Galois, Riemann. Fourth edition. 522pp. 5⅜ x 8½.
0-486-20630-0

THE HISTORY OF THE CALCULUS AND ITS CONCEPTUAL DEVELOP-MENT, Carl B. Boyer. Origins in antiquity, medieval contributions, work of Newton, Leibniz, rigorous formulation. Treatment is verbal. 346pp. 5⅜ x 8½. 0-486-60509-4

THE HISTORICAL ROOTS OF ELEMENTARY MATHEMATICS, Lucas N. H. Bunt, Phillip S. Jones, and Jack D. Bedient. Fundamental underpinnings of modern arithmetic, algebra, geometry and number systems derived from ancient civilizations. 320pp. 5⅜ x 8½.
0-486-25563-8

A HISTORY OF MATHEMATICAL NOTATIONS, Florian Cajori. This classic study notes the first appearance of a mathematical symbol and its origin, the competition it encountered, its spread among writers in different countries, its rise to popularity, its eventual decline or ultimate survival. Original 1929 two-volume edition presented here in one volume. xxviii+820pp. 5⅜ x 8½.
0-486-67766-4

GAMES, GODS & GAMBLING: A HISTORY OF PROBABILITY AND STATISTICAL IDEAS, F. N. David. Episodes from the lives of Galileo, Fermat, Pascal, and others illustrate this fascinating account of the roots of mathematics. Features thought-provoking references to classics, archaeology, biography, poetry. 1962 edition. 304pp. 5⅜ x 8½. (Available in U.S. only.)
0-486-40023-9

OF MEN AND NUMBERS: THE STORY OF THE GREAT MATHEMATICIANS, Jane Muir. Fascinating accounts of the lives and accomplishments of history's greatest mathematical minds–Pythagoras, Descartes, Euler, Pascal, Cantor, many more. Anecdotal, illuminating. 30 diagrams. Bibliography. 256pp. 5⅜ x 8½.
0-486-28973-7

HISTORY OF MATHEMATICS, David E. Smith. Nontechnical survey from ancient Greece and Orient to late 19th century; evolution of arithmetic, geometry, trigonometry, calculating devices, algebra, the calculus. 362 illustrations. 1,355pp. 5⅜ x 8½. Two-vol. set.        Vol. I: 0-486-20429-4   Vol. II: 0-486-20430-8

A CONCISE HISTORY OF MATHEMATICS, Dirk J. Struik. The best brief history of mathematics. Stresses origins and covers every major figure from ancient Near East to 19th century. 41 illustrations. 195pp. 5⅜ x 8½.        0-486-60255-9

# Physics

OPTICAL RESONANCE AND TWO-LEVEL ATOMS, L. Allen and J. H. Eberly. Clear, comprehensive introduction to basic principles behind all quantum optical resonance phenomena. 53 illustrations. Preface. Index. 256pp. 5⅜ x 8½. 0-486-65533-4

QUANTUM THEORY, David Bohm. This advanced undergraduate-level text presents the quantum theory in terms of qualitative and imaginative concepts, followed by specific applications worked out in mathematical detail. Preface. Index. 655pp. 5⅜ x 8½. 0-486-65969-0

ATOMIC PHYSICS (8th EDITION), Max Born. Nobel laureate's lucid treatment of kinetic theory of gases, elementary particles, nuclear atom, wave-corpuscles, atomic structure and spectral lines, much more. Over 40 appendices, bibliography. 495pp. 5⅜ x 8½. 0-486-65984-4

A SOPHISTICATE'S PRIMER OF RELATIVITY, P. W. Bridgman. Geared toward readers already acquainted with special relativity, this book transcends the view of theory as a working tool to answer natural questions: What is a frame of reference? What is a "law of nature"? What is the role of the "observer"? Extensive treatment, written in terms accessible to those without a scientific background. 1983 ed. xlviii+172pp. 5⅜ x 8½. 0-486-42549-5

AN INTRODUCTION TO HAMILTONIAN OPTICS, H. A. Buchdahl. Detailed account of the Hamiltonian treatment of aberration theory in geometrical optics. Many classes of optical systems defined in terms of the symmetries they possess. Problems with detailed solutions. 1970 edition. xv + 360pp. 5⅜ x 8½. 0-486-67597-1

PRIMER OF QUANTUM MECHANICS, Marvin Chester. Introductory text examines the classical quantum bead on a track: its state and representations; operator eigenvalues; harmonic oscillator and bound bead in a symmetric force field; and bead in a spherical shell. Other topics include spin, matrices, and the structure of quantum mechanics; the simplest atom; indistinguishable particles; and stationary-state perturbation theory. 1992 ed. xiv+314pp. 6⅛ x 9¼. 0-486-42878-8

LECTURES ON QUANTUM MECHANICS, Paul A. M. Dirac. Four concise, brilliant lectures on mathematical methods in quantum mechanics from Nobel Prize-winning quantum pioneer build on idea of visualizing quantum theory through the use of classical mechanics. 96pp. 5⅜ x 8½. 0-486-41713-1

THIRTY YEARS THAT SHOOK PHYSICS: THE STORY OF QUANTUM THEORY, George Gamow. Lucid, accessible introduction to influential theory of energy and matter. Careful explanations of Dirac's anti-particles, Bohr's model of the atom, much more. 12 plates. Numerous drawings. 240pp. 5⅜ x 8½. 0-486-24895-X

ELECTRONIC STRUCTURE AND THE PROPERTIES OF SOLIDS: THE PHYSICS OF THE CHEMICAL BOND, Walter A. Harrison. Innovative text offers basic understanding of the electronic structure of covalent and ionic solids, simple metals, transition metals and their compounds. Problems. 1980 edition. 582pp. 6⅛ x 9¼. 0-486-66021-4

# CATALOG OF DOVER BOOKS

HYDRODYNAMIC AND HYDROMAGNETIC STABILITY, S. Chandrasekhar. Lucid examination of the Rayleigh-Benard problem; clear coverage of the theory of instabilities causing convection. 704pp. 5⅜ x 8¼.                    0-486-64071-X

INVESTIGATIONS ON THE THEORY OF THE BROWNIAN MOVEMENT, Albert Einstein. Five papers (1905–8) investigating dynamics of Brownian motion and evolving elementary theory. Notes by R. Fürth. 122pp. 5⅜ x 8½. 0-486-60304-0

THE PHYSICS OF WAVES, William C. Elmore and Mark A. Heald. Unique overview of classical wave theory. Acoustics, optics, electromagnetic radiation, more. Ideal as classroom text or for self-study. Problems. 477pp. 5⅜ x 8½.      0-486-64926-1

GRAVITY, George Gamow. Distinguished physicist and teacher takes reader-friendly look at three scientists whose work unlocked many of the mysteries behind the laws of physics: Galileo, Newton, and Einstein. Most of the book focuses on Newton's ideas, with a concluding chapter on post-Einsteinian speculations concerning the relationship between gravity and other physical phenomena. 160pp. 5⅜ x 8½.
0-486-42563-0

PHYSICAL PRINCIPLES OF THE QUANTUM THEORY, Werner Heisenberg. Nobel Laureate discusses quantum theory, uncertainty, wave mechanics, work of Dirac, Schroedinger, Compton, Wilson, Einstein, etc. 184pp. 5⅜ x 8½. 0-486-60113-7

ATOMIC SPECTRA AND ATOMIC STRUCTURE, Gerhard Herzberg. One of best introductions; especially for specialist in other fields. Treatment is physical rather than mathematical. 80 illustrations. 257pp. 5⅜ x 8½.         0-486-60115-3

AN INTRODUCTION TO STATISTICAL THERMODYNAMICS, Terrell L. Hill. Excellent basic text offers wide-ranging coverage of quantum statistical mechanics, systems of interacting molecules, quantum statistics, more. 523pp. 5⅜ x 8½.
0-486-65242-4

THEORETICAL PHYSICS, Georg Joos, with Ira M. Freeman. Classic overview covers essential math, mechanics, electromagnetic theory, thermodynamics, quantum mechanics, nuclear physics, other topics. First paperback edition. xxiii + 885pp. 5⅜ x 8½.                                                                    0-486-65227-0

PROBLEMS AND SOLUTIONS IN QUANTUM CHEMISTRY AND PHYSICS, Charles S. Johnson, Jr. and Lee G. Pedersen. Unusually varied problems, detailed solutions in coverage of quantum mechanics, wave mechanics, angular momentum, molecular spectroscopy, more. 280 problems plus 139 supplementary exercises. 430pp. 6½ x 9¼.                                      0-486-65236-X

THEORETICAL SOLID STATE PHYSICS, Vol. 1: Perfect Lattices in Equilibrium; Vol. II: Non-Equilibrium and Disorder, William Jones and Norman H. March. Monumental reference work covers fundamental theory of equilibrium properties of perfect crystalline solids, non-equilibrium properties, defects and disordered systems. Appendices. Problems. Preface. Diagrams. Index. Bibliography. Total of 1,301pp. 5⅜ x 8½. Two volumes.          Vol. I: 0-486-65015-4   Vol. II: 0-486-65016-2

WHAT IS RELATIVITY? L. D. Landau and G. B. Rumer. Written by a Nobel Prize physicist and his distinguished colleague, this compelling book explains the special theory of relativity to readers with no scientific background, using such familiar objects as trains, rulers, and clocks. 1960 ed. vi+72pp. 5⅜ x 8½.       0-486-42806-0

## *CATALOG OF DOVER BOOKS*

A TREATISE ON ELECTRICITY AND MAGNETISM, James Clerk Maxwell. Important foundation work of modern physics. Brings to final form Maxwell's theory of electromagnetism and rigorously derives his general equations of field theory. 1,084pp. 5⅜ x 8½. Two-vol. set. Vol. I: 0-486-60636-8 Vol. II: 0-486-60637-6

QUANTUM MECHANICS: PRINCIPLES AND FORMALISM, Roy McWeeny. Graduate student-oriented volume develops subject as fundamental discipline, opening with review of origins of Schrödinger's equations and vector spaces. Focusing on main principles of quantum mechanics and their immediate consequences, it concludes with final generalizations covering alternative "languages" or representations. 1972 ed. 15 figures. xi+155pp. 5⅜ x 8½. 0-486-42829-X

INTRODUCTION TO QUANTUM MECHANICS With Applications to Chemistry, Linus Pauling & E. Bright Wilson, Jr. Classic undergraduate text by Nobel Prize winner applies quantum mechanics to chemical and physical problems. Numerous tables and figures enhance the text. Chapter bibliographies. Appendices. Index. 468pp. 5⅜ x 8½. 0-486-64871-0

METHODS OF THERMODYNAMICS, Howard Reiss. Outstanding text focuses on physical technique of thermodynamics, typical problem areas of understanding, and significance and use of thermodynamic potential. 1965 edition. 238pp. 5⅜ x 8½. 0-486-69445-3

THE ELECTROMAGNETIC FIELD, Albert Shadowitz. Comprehensive undergraduate text covers basics of electric and magnetic fields, builds up to electromagnetic theory. Also related topics, including relativity. Over 900 problems. 768pp. 5⅝ x 8¼. 0-486-65660-8

GREAT EXPERIMENTS IN PHYSICS: FIRSTHAND ACCOUNTS FROM GALILEO TO EINSTEIN, Morris H. Shamos (ed.). 25 crucial discoveries: Newton's laws of motion, Chadwick's study of the neutron, Hertz on electromagnetic waves, more. Original accounts clearly annotated. 370pp. 5⅜ x 8½. 0-486-25346-5

EINSTEIN'S LEGACY, Julian Schwinger. A Nobel Laureate relates fascinating story of Einstein and development of relativity theory in well-illustrated, nontechnical volume. Subjects include meaning of time, paradoxes of space travel, gravity and its effect on light, non-Euclidean geometry and curving of space-time, impact of radio astronomy and space-age discoveries, and more. 189 b/w illustrations. xiv+250pp. 8⅜ x 9¼. 0-486-41974-6

STATISTICAL PHYSICS, Gregory H. Wannier. Classic text combines thermodynamics, statistical mechanics and kinetic theory in one unified presentation of thermal physics. Problems with solutions. Bibliography. 532pp. 5⅜ x 8½. 0-486-65401-X

Paperbound unless otherwise indicated. Available at your book dealer, online at **www.doverpublications.com**, or by writing to Dept. GI, Dover Publications, Inc., 31 East 2nd Street, Mineola, NY 11501. For current price information or for free catalogues (please indicate field of interest), write to Dover Publications or log on to **www.doverpublications.com** and see every Dover book in print. Dover publishes more than 500 books each year on science, elementary and advanced mathematics, biology, music, art, literary history, social sciences, and other areas.

## *CATALOG OF DOVER BOOKS*

TENSOR CALCULUS, J.L. Synge and A. Schild. Widely used introductory text covers spaces and tensors, basic operations in Riemannian space, non-Riemannian spaces, etc. 324pp. 5⅜ x 8¼. 0-486-63612-7

ORDINARY DIFFERENTIAL EQUATIONS, Morris Tenenbaum and Harry Pollard. Exhaustive survey of ordinary differential equations for undergraduates in mathematics, engineering, science. Thorough analysis of theorems. Diagrams. Bibliography. Index. 818pp. 5⅜ x 8½. 0-486-64940-7

INTEGRAL EQUATIONS, F. G. Tricomi. Authoritative, well-written treatment of extremely useful mathematical tool with wide applications. Volterra Equations, Fredholm Equations, much more. Advanced undergraduate to graduate level. Exercises. Bibliography. 238pp. 5⅜ x 8½. 0-486-64828-1

FOURIER SERIES, Georgi P. Tolstov. Translated by Richard A. Silverman. A valuable addition to the literature on the subject, moving clearly from subject to subject and theorem to theorem. 107 problems, answers. 336pp. 5⅜ x 8½. 0-486-63317-9

INTRODUCTION TO MATHEMATICAL THINKING, Friedrich Waismann. Examinations of arithmetic, geometry, and theory of integers; rational and natural numbers; complete induction; limit and point of accumulation; remarkable curves; complex and hypercomplex numbers, more. 1959 ed. 27 figures. xii+260pp. 5⅜ x 8½. 0-486-63317-9

POPULAR LECTURES ON MATHEMATICAL LOGIC, Hao Wang. Noted logician's lucid treatment of historical developments, set theory, model theory, recursion theory and constructivism, proof theory, more. 3 appendixes. Bibliography. 1981 edition. ix + 283pp. 5⅜ x 8½. 0-486-67632-3

CALCULUS OF VARIATIONS, Robert Weinstock. Basic introduction covering isoperimetric problems, theory of elasticity, quantum mechanics, electrostatics, etc. Exercises throughout. 326pp. 5⅜ x 8½. 0-486-63069-2

THE CONTINUUM: A CRITICAL EXAMINATION OF THE FOUNDATION OF ANALYSIS, Hermann Weyl. Classic of 20th-century foundational research deals with the conceptual problem posed by the continuum. 156pp. 5⅜ x 8½. 0-486-67982-9

CHALLENGING MATHEMATICAL PROBLEMS WITH ELEMENTARY SOLUTIONS, A. M. Yaglom and I. M. Yaglom. Over 170 challenging problems on probability theory, combinatorial analysis, points and lines, topology, convex polygons, many other topics. Solutions. Total of 445pp. 5⅜ x 8½. Two-vol. set.
Vol. I: 0-486-65536-9 Vol. II: 0-486-65537-7

Paperbound unless otherwise indicated. Available at your book dealer, online at **www.doverpublications.com**, or by writing to Dept. GI, Dover Publications, Inc., 31 East 2nd Street, Mineola, NY 11501. For current price information or for free catalogues (please indicate field of interest), write to Dover Publications or log on to **www.doverpublications.com** and see every Dover book in print. Dover publishes more than 500 books each year on science, elementary and advanced mathematics, biology, music, art, literary history, social sciences, and other areas.